CREATIV
DO IT YOURS

Wiring and Lighting

WARD LOCK

© Ward Lock Limited, 1994
A Cassell Imprint
Villiers House, 41-47 Strand, London WC2N 5JE

Based on *Successful DIY*
© Eaglemoss Publications Limited, 1994

ISBN 0 7063 7280 8

Printed in Spain by Cayfosa Industria Grafica

10 9 8 7 6 5 4 3 2 1

CONTENTS

INTRODUCTION

MANY do-it-yourselfers almost never touch electrical work because they are afraid of doing something wrong and causing injury or an accident. While it is true that electricity can kill if it is not treated with respect, so long as you know what you are doing there is little to fear, since wiring work actually involves less skill than many other jobs in the home.

If you are attempting wiring work for the first time, the first part of this book will help you familiarize yourself with the electrical system in your home starting at the consumer unit and working round the circuits to the lights, the power points and the various individual appliances – cookers, showers, immersion heaters and so on – which have their own circuits. It uses clear, coloured wiring diagrams to explain how individual wiring accessories are connected up, how the separate circuits are run and how to test a circuit if you have to trace a fault. It also describes cable and flex – the raw materials used for wiring work – to help you choose the right type and rating for every job.

Electrical repairs come next, with detailed step-by-step instructions on how to wire a new plug, replace worn flex and repair blown fuses in both plugs and fuse boxes – in the latter situation, with the help of a helpful fault-finder flow chart. This section also explains how to replace damaged wiring accessories, which could expose live parts if not attended to.

The rest of the book covers a range of projects for extending existing wiring and making it more versatile. Two of the most useful are adding extra power points, which will enable you to avoid using adaptors and extension flexes, and fitting a special socket outlet for an electric shaver.

Lighting comes next, with instructions for providing two-way switching, installing a new pendant light, and fitting wall lights. If you want some more unusual lighting effects, the book explains how to add spotlights, downlighters and track lighting and how to replace ordinary switches with dimmers to give you complete control of light levels. There is also an invaluable section on fitting security lighting outdoors – one of the most effective ways of protecting your property against intruders.

The final section deals with two related projects which do not involve mains wiring, but which demand similar skills and techniques: wiring additional telephone sockets, and adding extra TV aerial sockets.

Throughout the book the emphasis is on safety, with detailed shopping lists, clear wiring diagrams and step-by-step explanations for every job. All the instructions given comply with the Wiring Regulations, and will enable you to carry out your own wiring work with confidence and to professional standards.

UNDERSTANDING ELECTRICITY

Light, warmth, entertainment and a helping hand all at the flick of a switch – and all thanks to electricity, the power at our fingertips. Yet, despite its enormous advantages, electricity can be a deadly force if it is not treated with respect. So even if you do not plan to carry out any electrical work in your home yourself, it is important to gain a basic understanding of how the system works. As a consumer, there are certain facts which it pays to know – both for safety reasons and so that you can be sure of getting value for money when you arrange for work to be done.

Who supplies electricity?
Currently, electricity supply in the United Kingdom is the responsibility of electricity supply companies. Electricity is generated by two companies – called Powergen and National Power – and is then distributed via a system of power lines called the National Grid.

From there power is then transferred by the various supply companies to their own local substations. From here, it travels along underground or overhead mains to individual homes in the area.

On arrival, the supply runs to the meter which gives the supply company an accurate reading of how much electricity the household has consumed over a given period. There are various different types of meter to suit different payment methods (see below), but all of them remain the property of the supply company.

Your responsibilities
As with other services, responsibility for the electrical system from the meter onwards rests with the householder. You are obliged by law to keep it in a good state of repair, and the supply company can disconnect the supply if they know it to be dangerous. Alternatively, they can serve you with a notice to carry out repairs within a certain time and then check that you have done so.

The supply company will also check out your system on request, although you have to pay a fee for this service.

YOU AND YOUR METER
There are four types of domestic electricity meter, and you can ask the supply company to install whichever type best suits your needs.

Domestic standard meters charge at the standard rate. Modern versions have digital read-outs in place of the old six-dial system.

Pre-payment meters are coin-operated; you pay for the electricity before you use it.

Budget meters are similar, but are operated by a special key. The key is rechargeable with set numbers of pre-paid units at the supply company's showrooms.

Economy 7 meters allow you to take advantage of electricity at a reduced off-peak rate.

Meter replacement
The supply companies are obliged to replace meters every 20 years at their own expense. If you receive a notice to this effect, you are obliged by law to comply with it, and the company's engineers can insist that the meter is made readily accessible. Householders are liable for meters damaged through their own negligence.

Like gas meters, electricity meters can now be relocated inside weatherproof boxes out of doors so that they are available for reading at any time. The box is locked and both meter reader and householder have a key. Unlike British Gas, the electricity supply companies don't as yet have a policy of changing entire areas to outdoor meters, but they are available on request.

An outdoor meter is a good idea if there's no one at home all day.

HOUSEHOLD ELECTRICAL SYSTEMS

Household electrical systems consist of a number of separate *circuits*. Some supply power socket outlets, others supply the fixed lighting, and there are separate circuits for individual high-power appliances such as cookers and showers.

Each circuit starts at the consumer unit. It is rated according to the total amount of current it can safely carry, and the fuse or MCB in the consumer unit which protects it is sized accordingly (see opposite).

Power circuits for socket outlets can either be wired in strings (called *radial circuits*) or in continuous chains (called *ring circuits*). In both cases, *spurs* may be taken off the main circuits to supply extra outlets. Power circuits for single appliances are always wired on the radial system.

Lighting circuits are radial too, but can be wired via *junction boxes* or *looped in* directly to the light fittings.

All the circuits are wired using flat-section *PVC-sheathed cable*.

Like circuit fuses, cable is sized according to the current rating of the circuit it supplies – the more current it has to take, the thicker it is.

Where possible, the cable is usually hidden – under floors, in the loft, through the cavities in stud walls, or through channels (chases) in the wall plaster. But cable can also be surface-mounted, in which case it is either clipped along the tops of skirtings and architraves, or run in impact-resistant plastic mini-trunking.

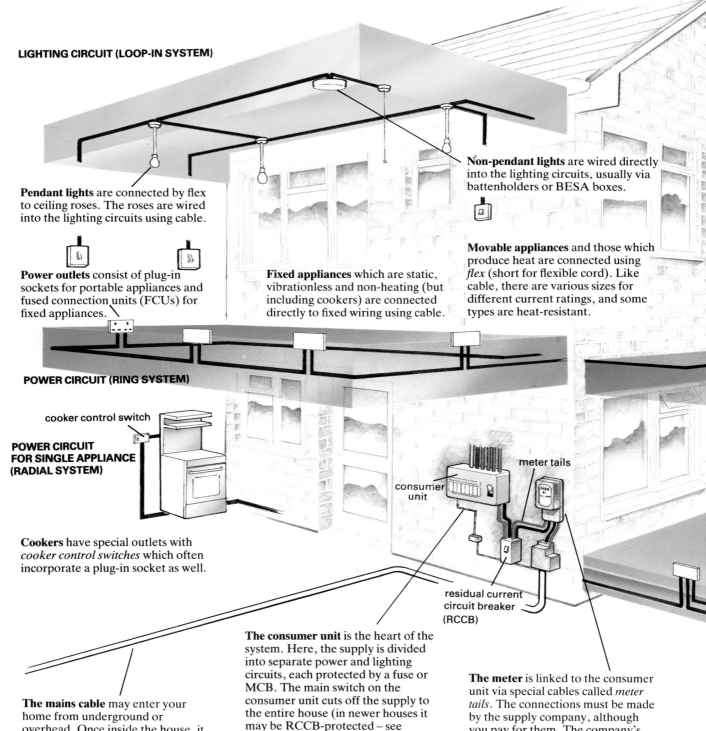

LIGHTING CIRCUIT (LOOP-IN SYSTEM)

Pendant lights are connected by flex to ceiling roses. The roses are wired into the lighting circuits using cable.

Power outlets consist of plug-in sockets for portable appliances and fused connection units (FCUs) for fixed appliances.

Fixed appliances which are static, vibrationless and non-heating (but including cookers) are connected directly to fixed wiring using cable.

Non-pendant lights are wired directly into the lighting circuits, usually via battenholders or BESA boxes.

Movable appliances and those which produce heat are connected using *flex* (short for flexible cord). Like cable, there are various sizes for different current ratings, and some types are heat-resistant.

POWER CIRCUIT (RING SYSTEM)

cooker control switch

POWER CIRCUIT FOR SINGLE APPLIANCE (RADIAL SYSTEM)

meter tails

consumer unit

Cookers have special outlets with *cooker control switches* which often incorporate a plug-in socket as well.

residual current circuit breaker (RCCB)

The mains cable may enter your home from underground or overhead. Once inside the house, it runs via the supply company's sealed fuse to the meter.

The consumer unit is the heart of the system. Here, the supply is divided into separate power and lighting circuits, each protected by a fuse or MCB. The main switch on the consumer unit cuts off the supply to the entire house (in newer houses it may be RCCB-protected – see opposite).

The meter is linked to the consumer unit via special cables called *meter tails*. The connections must be made by the supply company, although you pay for them. The company's responsibility for the system ends at the tails.

VOLTS, AMPS, WATTS

If you think of electricity in a wire as water in a pipe:

Volts measure voltage – the pressure or 'push' exerted on the electricity; in British homes, it is pushed through at 240 volts.

Amps measure current – the amount of electricity available for use; British consumer units are rated at up to 100 amps, and an average house with all its electrical appliances switched on won't normally demand more than this.

Watts measure power – the rate at which electrical energy is used. On high power appliances, **kilowatts** (each of which is 1,000 watts) are used instead.

For metering purposes, the amount is charged on the basis of kilowatt hours – the total number of kilowatts consumed multiplied by the hours during which they were used.

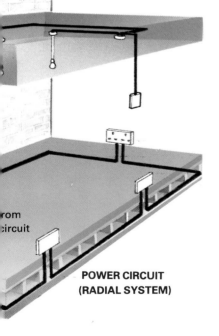

LIGHTING CIRCUIT (JUNCTION BOX SYSTEM)

from circuit

POWER CIRCUIT (RADIAL SYSTEM)

USING ELECTRICITY SAFELY

British-style electrical systems are among the safest in the world. All the components – from the cable and flex used for the wiring, to sockets, plugs and light fittings – are designed with a large built-in safety margin. In addition:

■ All domestic systems should be fully earthed, which means that any current that escapes during a fault passes to earth, causing the supply to be immediately shut off.

■ All houses have a two-stage circuit breaking system to cut off the supply in the event of a fault. Plug fuses, which protect individual appliances and their connections, form the first stage; circuit fuses or miniature circuit breakers (MCBs), which protect the house's fixed wiring circuits, form the second stage.

In newer installations, there is a third stage – the Residual Current Circuit Breaker (RCCB) – which shuts off part or all of the supply (depending on the type) in the event of a fault.

Electrocution is not the only danger – a faulty or overloaded electrical system can generate heat on a massive scale, creating a potential fire hazard. For safety's sake, follow these simple rules:

Check all plugs are properly and securely wired, and that they are fitted with the correct fuses.

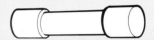

Avoid trailing flexes – they are easy to trip over or damage. Don't coil up flexes, and always unroll extension leads completely; coils can generate enough heat to melt the insulation and start a fire.

Check all flexes at least once a year for signs of damage. Have any flexes patched with insulating tape replaced, and don't use taped joints to extend flexes.

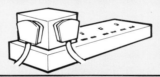

Avoid adaptors and permanent extension leads, both of which are easily overloaded. Install extra power sockets instead, or convert single sockets to double ones.

Use a plug-in RCCB with all outdoor appliances. If you cut through the lead by mistake, turn off the supply before touching it – even if there is an RCCB.

Never place anything wet (including your hands) near an electrical appliance, bulb, light fitting, switch or power outlet.

Never by-pass fuses which have blown (or tape down tripped miniature circuit breakers). Always replace the fuse with correct size wire or cartridge.

And if you plan to do electrical work yourself . . .

Switch off at the main switch on the consumer unit before tampering in any way with the fixed circuit wiring (including light fittings and power points).

If you work on the wiring take the following precautions:

■ First find out which circuit you are dealing with.

■ Remove the relevant fuse or tape MCB in 'off' position. Keep the fuse with you so that no-one else can replace it without you knowing.

■ Restore the power and check that the circuit is 'dead'.

Call an electrician if you are in any doubt about which wire goes where.

ELECTRICAL WORK AND THE LAW

In England, Wales and N. Ireland there are no laws as such to stop you from doing electrical work of any kind. However, all professional electricians follow the Institution of Electrical Engineers' (IEE) Wiring Regulations (see below), and amateurs are strongly advised to do likewise – the Regulations exist, after all, for everyone's protection.

Although you can't be prosecuted for ignoring the Wiring Regulations, an installation which doesn't conform can be condemned and cut off by the supply company.

In Scotland, the Wiring Regulations are incorporated in the Scottish Building Regulations. This gives them the force of law, and you can be prosecuted directly for disobeying them.

What the Regulations cover

Copies of the latest edition of the Wiring Regulations (they are regularly updated) can be found at your local library. Unfortunately, such is their complexity that they can be very difficult for the layman to interpret. The chart on the right summarises the areas they cover.

Fittings and materials	■ The Regulations specify how electrical parts are made and what they are made of.
Rating of circuits	■ Circuits are current-rated according to the maximum load they can handle. Cables are sized and fuses rated accordingly.
Cable runs	■ The Regulations specify how and where cables can be run. There are also charts showing how a cable's rating is affected by the surrounding temperature, thermal insulation, or the presence of other cables.
Positioning of fittings	■ There are restrictions on the placement of sockets, designed to guard against accidental damage or contact with water. ■ Cooker control switches have to be readily accessible. ■ Plug-in sockets and portable appliances are effectively banned from bathrooms. Similarly, light switches must be the pull-cord type, and bare bulbs must be enclosed to guard against moisture.
Size of circuits	■ There are restrictions on the total floor area served by circuits, depending on the type of circuit, the size of cable used, and what type and rating of fuse is fitted.
Cross-bonding	■ All incoming/outgoing metal pipes and fittings in new or rewired installations have to be cross-bonded – connected to the main earth terminal at the consumer unit – so that they are earthed should they come into contact with a live wire. ■ In bathrooms, any accessible metalwork which could conceivably bring in electrical current from outside must be cross-bonded.
Quality of work	■ There are guidelines on how and where to make connections, and what parts to use.
	NB All instructions in this book comply with the relevant IEE Wiring Regulations.

MAKING IMPROVEMENTS

By far the most important improvement is to rewire an installation which is nearing the end of its useful life.

Before PVC sheathed cable was introduced in the early 1950s, homes were wired using *rubber sheathed cable*. This has a lifespan of only 25 years, after which it starts to break down and become a progressively greater fire risk, so any home still containing it is now long overdue for rewiring.

An early PVC cable-wired installation may be suspect too, since heat and certain chemicals can cause the sheathing to break down – particularly around light fittings, switches and sockets. It's also possible for the system to have been partially rewired leaving some parts decidedly unsafe.

So if you've just moved, or you are unsure how old the existing wiring is, have the installation checked professionally as soon as possible – even if you have had a structural survey.

Other improvements

Adding an extra socket here and there is usually relatively easy to arrange, although running new cables can damage existing decorations. But major additions could require extra circuits, in which case your consumer unit may not be large enough to cope.

One answer is to install a smaller *auxiliary consumer unit* (often called a 'switchfuse unit'), but a better investment in the long term is to have the existing consumer unit replaced with a larger one incorporating MCBs instead of fuses.

Older houses may still have obsolete wiring which must be replaced with PVC sheathed cable.

UNDERSTANDING LIGHTING CIRCUITS

Understanding how house lighting circuits work makes it easier to track down faults, and is essential if you plan to alter or extend your lighting in the future.

The previous section explained how to identify the fuses or miniature circuit breakers (MCBs) in the consumer unit which protect your lighting circuits. There are probably two of them – one for the upstairs circuit, and one for the downstairs – unless you live in a small flat, in which case there may be only one.

Each fuse or MCB will be rated at 5 amps, and may carry a white dot for identification. If you look at their holders in the consumer unit, you will see that from here there are main circuit cables running to all the lights in the house.

Safety first!
Always turn off the power at the main switch on the consumer unit and pull out the lighting circuit fuses (or turn off the MCBs) before touching any part of your lighting circuits.

TWO THINGS TO REMEMBER

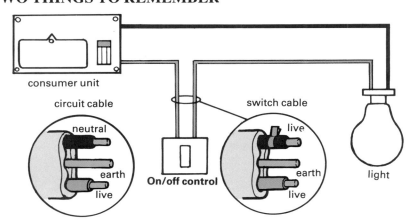

consumer unit

circuit cable · switch cable

neutral · earth · live

On/off control

live · earth · live

light

Two things about lighting circuits confuse a lot of people:

1 On/off control In all lighting circuits the current flows out from the consumer unit along the live (red) cable core, and back along the neutral (black) core. In between, it's intercepted by a switch which breaks the flow as shown in the diagram above.

Confusingly, the same cable is used for the switch connections as is used for the rest of the circuit. This means the current flows into the switch along the red core and back out again along the black core – so both are in fact live. In practice the black core on a switch cable should be marked with a piece of red tape to remind anyone that this is so, though often it isn't.

2 Two wiring systems All lighting circuits are wired up as what are known as radial circuits – the cable supplies each light in turn, as far as the one furthest from the consumer unit.

There are two ways of wiring the cable, as shown left. You may find one or the other, or both.
Loop-in wiring is the modern system. The cable runs direct to each light fitting, which also contains the switch connections.
Junction box wiring is the old system. The circuit cables are connected to a series of junction boxes, and further cables run from these to each light and switch.

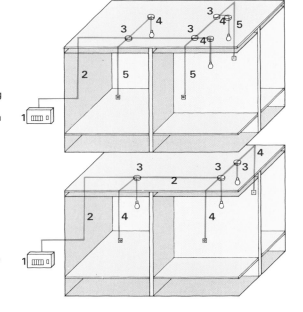

KEY

Junction box wiring

1 consumer unit
2 circuit cable
3 junction box
4 connection to ceiling rose and light
5 connection to switch

KEY

Loop-in wiring

1 consumer unit
2 circuit cable
3 ceiling rose
4 connection to switch

YOUR LIGHTS UPSTAIRS

The first step in discovering how your lighting circuits are wired up is to draw sketch plans of each floor of the house, starting with the upstairs as shown on the right. Mark in the positions of every switch and light, so you have a record of what controls what.

Now check which fuse or MCB controls the upstairs circuit:
■ Switch off at the mains and remove one of the lighting fuses (or trip one of the MCBs).
■ Switch on again and see which lights remain off. If you've chosen the downstairs one, repeat the experiment with the other lighting fuse or MCB just to make sure. Record your findings and label the fuses.

Switch the mains **OFF** again. You can now open up the various fittings to see how they're wired, referring to the examples below. It will be helpful if you can get into the loft, since this is just about the only place where lighting cables are exposed. Record all your findings on the plan.

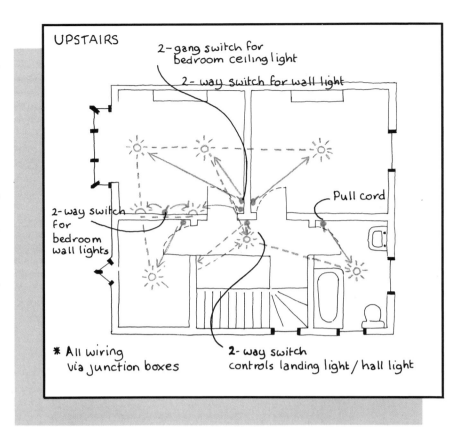

UPSTAIRS

2-gang switch for bedroom ceiling light

2-way switch for wall light

2-way switch for bedroom wall lights

Pull cord

* All wiring via junction boxes

2-way switch controls landing light / hall light

JUNCTION BOX WIRING

With the power off, unscrew the cover of each **ceiling rose** or **battenholder** in turn. If only one cable is present, you know the fitting is wired from a **junction box**. This box links the circuit cable to the light fitting and light switch. It contains four cables unless it is the last on the circuit, in which case there are only three.

The flex linking the rose to the **lampholder** has just two cores – unless the lampholder is metal, in which case it *should* have three-core flex incorporating an earth core.

Next, open up any **wall switches** and bathroom **ceiling switches**. Unless they are wired up for multi-way switching (see overleaf), there will be only one cable. With plastic switches, the earth core should be connected to a terminal on the mounting box.

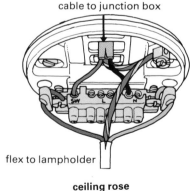

cable to junction box

flex to lampholder

ceiling rose

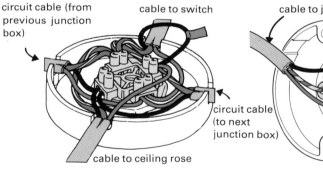

circuit cable (from previous junction box)

cable to switch

circuit cable (to next junction box)

cable to ceiling rose

junction box

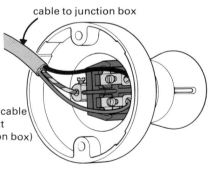

cable to junction box

battenholder

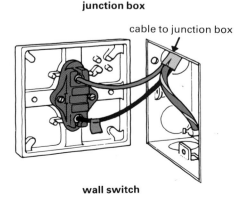

metal lampholder

cable to junction box

wall switch

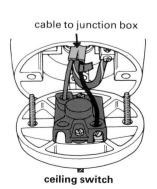

cable to junction box

ceiling switch

YOUR LIGHTS DOWNSTAIRS

Make a plan for the downstairs lighting as you did for the upstairs, following the example on the right. Working out the wiring is likely to be more difficult since the cables are concealed between the ceiling and the floor above, but by this stage you should have a clearer idea of how things work.

As you progress, you're likely to find some extra features.

Multi-way switching: The switch in the hall is usually wired to control both the hall and landing lights. See overleaf.

Wall lights: Each one may have its own switch, or the lights may be controlled as a group from a single switch – in which case they'll be wired to it via a junction box.

It's also possible for wall lights not to be on a lighting circuit at all (which you will discover when you check the fuses). Sometimes they're fed from individually fused connections taken from a power circuit.

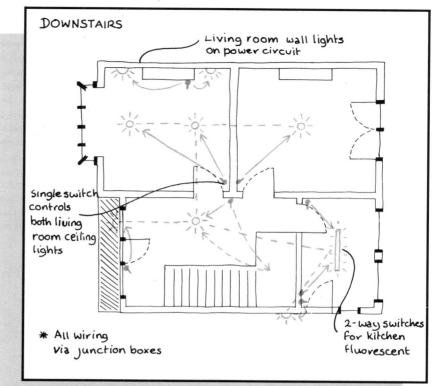

DOWNSTAIRS

Living room wall lights on power circuit

Single switch controls both living room ceiling lights

* All wiring via junction boxes

2-way switches for kitchen fluorescent

LOOP-IN WIRING

If you find two or more cables when you open up a ceiling rose, this indicates that the fitting is wired up on the loop-in system.

A **loop-in rose** has three groups of terminals – two for the circuit cables linking it to the fitting before and the fitting after; and one for the switch cable. (If the rose is the last in line on the system, it has only two cables.)

Plastic lampholders don't need earthing and are connected to the rose with two core flex.

Wall and ceiling switches are wired as in a junction box system.

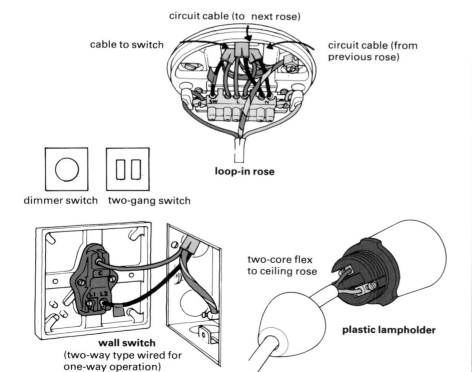

circuit cable (to next rose)

cable to switch

circuit cable (from previous rose)

loop-in rose

dimmer switch   two-gang switch

two-core flex to ceiling rose

plastic lampholder

wall switch
(two-way type wired for one-way operation)

SWITCH VARIATIONS

Don't let these different types of switch confuse you:

Two-way switches (see overleaf) are sometimes wired for one-way operation, in which case one of the terminals is left empty as shown on the left.

Two- and three-gang switches control two or three lights from the same place. They simply have two (or three) separate switch cables instead of one.

Dimmer switches look different, but are wired in exactly the same way as other switches.

FUSED SPURS

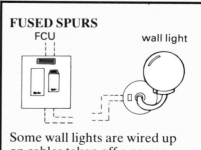

FCU

wall light

Some wall lights are wired up on cables taken off a power circuit. The cable runs to a fused connection unit (FCU) containing a 5 amp fuse, and from there to the light itself. The connection unit may serve as the on/off switch too.

MULTI-WAY SWITCHING

Common examples of multi-way switching are between the hall and landing, and in a through-room with switches at both doors. Other possible places include corridors served by two lights, and bedroom wall lights with a switch by the door and another by the bed.

Lights wired in this way are controlled by special switches with three terminals instead of two.

In a basic two switches/one light arrangement, the cable from the loop-in rose or junction box runs to Switch 1, where it is connected to the terminals marked L1 and L2. Switch 1 is then connected to Switch 2 using special three-core and earth cable which links all three terminals together as shown right. Either switch can now be used to turn the light on and off.

More complex arrangements make use of two-gang switches and linked switch cables to provide inter-changeable control of two or more lights at once. It's also possible, though not common, to have three-way switching.

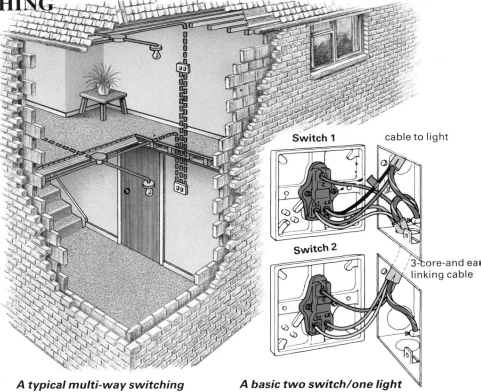

Switch 1

cable to light

Switch 2

3-core-and-ear linking cable

A typical multi-way switching arrangement in which both lights are controlled independently from either upstairs or downstairs using two-gang switches.

A basic two switch/one light arrangement, with the second switch linked to the first using 3-core cable so that both control the same light.

CLOSE-FITTING LIGHTS

Not all fixed light fittings are pendant lampholders or battenholders. Many are fixed directly to the ceiling, and since there is no pendant flex the connections have to be made in a different way.

Most close-fitting lights have a short length of flex emerging from their baseplates which is connected to the circuit cable using screw-down plastic block connectors.

UK Wiring Regulations require the connections to be housed in a special enclosure called a BESA box (conduit box). This is recessed into the ceiling and covered by the baseplate of the fitting.

The way the fittings are linked depends on whether loop-in or junction box wiring is being used, as shown below.

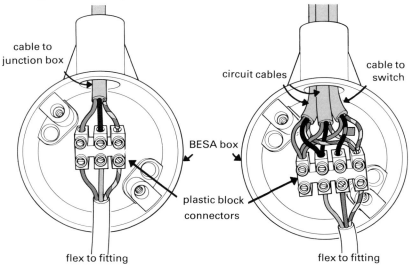

cable to junction box

BESA box

plastic block connectors

flex to fitting

circuit cables

cable to switch

flex to fitting

Junction box-wired BESA box with a single cable running to the junction box.

Loop-in-wired BESA box with cables running to the switch, the next fitting and the previous one.

FLUORESCENT LIGHTS

Fluorescent fittings – and most track lights – have fixed terminals somewhere on their baseplate to which the circuit cable is connected direct (remove the bulb and unscrew the cover to expose them). The terminals should be labelled L (live), N (neutral) and E (earth).

The cable runs from the fitting to a junction box, where it is linked to the switch cable – there is no facility for loop-in wiring at the fitting itself. Some types of wall light are also wired in a similar way.

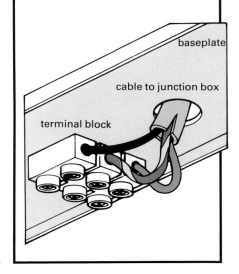

baseplate

cable to junction box

terminal block

UNDERSTANDING POWER CIRCUITS

Power circuits supply current to the sockets into which you plug your electrical appliances and lamps.

Some appliances in more or less constant use (ie waste disposers, washing machines) may not plug in, but instead connect directly to the power circuit. There are also special circuits for individual appliances which use a lot of electricity, such as cookers, showers and immersion heaters.

Like lighting circuits, the average house has two power circuits – one for upstairs, one for downstairs. They are wired using two-core-and-earth cable, which has one red core (live), one black core (neutral) and one bare copper core (earth). The earth core should be insulated with slip-on green and yellow PVC sheathing where it is exposed.

Both power circuits start at the consumer unit, and each has its own fuse and fuseholder or MCB (miniature circuit breaker). The fuses are usually 30 amps (red dot) but sometimes 20 amps (yellow dot).

From here, the circuits may be wired in one of two ways.

Types of power circuit

Radial circuit wiring is the old method. The cable runs from the consumer unit to each socket in turn in a 'daisy chain'. In other words, electricity flows out along the live side of the circuit, supplies each socket in the chain, and then flows back along the neutral side.

The problem with radial circuits is that the current can only travel in one direction, so if there are too many links in the chain the cable is overloaded. There are regulations designed to avoid this, but the system is wide open to abuse.

Ring circuit wiring is the modern method. The cable runs from the consumer unit to each socket on the circuit *and then back to the consumer unit again*. This gives the current a choice of two live wires to travel along when flowing to any socket – and similarly, a choice of two neutral wires to travel back along. Ring circuits therefore carry less load, which makes them safer.

Check the fuseholders in the consumer unit (right) to see which way your power circuits are wired.

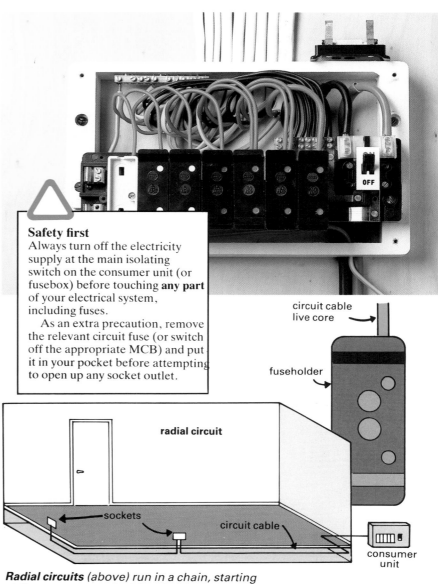

Safety first
Always turn off the electricity supply at the main isolating switch on the consumer unit (or fusebox) before touching **any part** of your electrical system, including fuses.

As an extra precaution, remove the relevant circuit fuse (or switch off the appropriate MCB) and put it in your pocket before attempting to open up any socket outlet.

Radial circuits (above) run in a chain, starting at the consumer unit. A fuseholder for a radial circuit (inset) has only one red cable core running into it – shut off the main switch and remove the fuse to check.

Ring ciruits (below) run in a loop, starting and ending at the consumer unit. A fuseholder for a ring circuit (inset) has two live cores running into it. Consumer units with MCBs are nearly always ring circuit-wired.

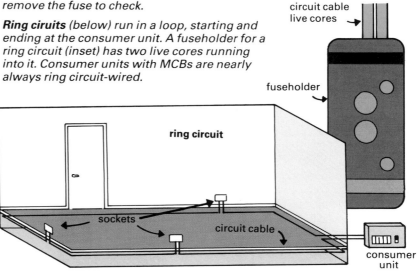

YOUR SOCKETS UPSTAIRS

It's easier to work out which power circuits control what if you draw sketch plans of your socket layouts as shown on the right.

Start upstairs. Mark on all the plug-in sockets, and also where any appliances are wired in directly (these are called fused connection units [FCUs] – see overleaf).

If you don't yet know which circuit fuse or MCB controls the upstairs power circuit, check now. Turn off the main switch at the consumer unit, remove one of the 30A fuses (or trip the equivalent MCB), switch on again, then check which sockets work.

Repeat the procedure, switching off each time, until you find the right fuse. Label it 'upstairs power'.

Now switch off, remove this fuse, restore the power again and check **every** upstairs socket and FCU. If any still work, you know they are wired from another circuit (probably the downstairs one). It's important to note this on your plan for future reference.

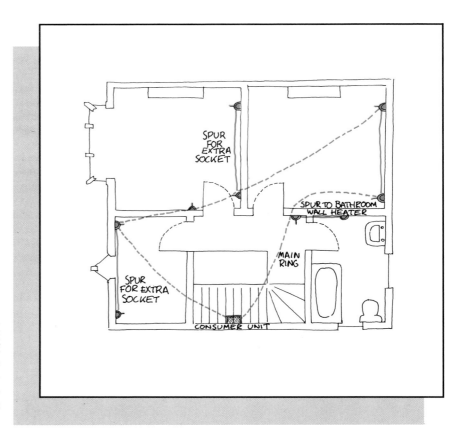

CHECKING SOCKETS

You can find out more about your power circuits by checking how many cables each socket contains and then referring to the diagrams.

First, turn the main switch at the consumer unit to OFF. Then unscrew each socket faceplate in turn and pull it gently away from its box.

As the diagrams show, it's possible to have **spur** sockets wired as branches off the main ring or radial circuit. There are strict regulations governing how many sockets can be wired in this way.

In some cases you can't tell from one socket alone whether it's ring or radially wired. Check by having another look at the fuseholder controlling that socket's circuit.

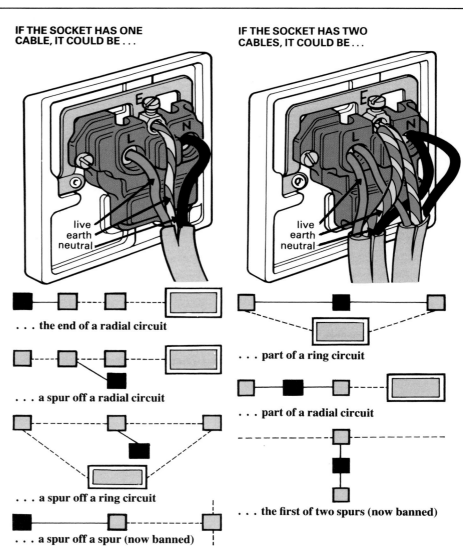

IF THE SOCKET HAS ONE CABLE, IT COULD BE . . .

. . . **the end of a radial circuit**

. . . **a spur off a radial circuit**

. . . **a spur off a ring circuit**

. . . **a spur off a spur (now banned)**

IF THE SOCKET HAS TWO CABLES, IT COULD BE . . .

. . . **part of a ring circuit**

. . . **part of a radial circuit**

. . . **the first of two spurs (now banned)**

YOUR SOCKETS DOWNSTAIRS

Follow the same procedure for the downstairs sockets and FCUs, again drawing a plan of their positions.

Downstairs, there are sometimes two power circuits: one for the kitchen (where the heaviest load occurs) and one for the other rooms. The kitchen circuit is likely to be a radial one, even if the others are ring-wired.

Bear this in mind when checking which fuse controls what. Remove each 30A fuse in turn (including the upstairs one), see which sockets and FCUs stop working, and tick them off on your plan. Label the power circuit fuses accordingly, and make a mental note that any which are left must control special circuits.

During your investigations, you may find one or more sockets controlled by the upstairs circuit fuse. This will probably have been done to save cable.

Finish by marking on the plans which sockets are served by which circuits.

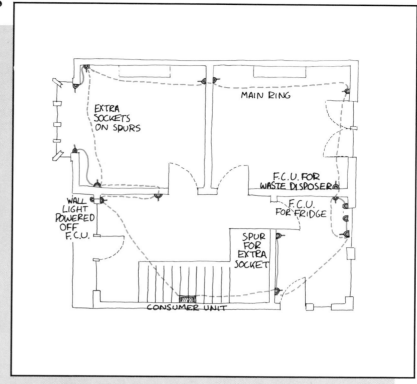

IF THE SOCKET HAS THREE CABLES, IT COULD BE . . .

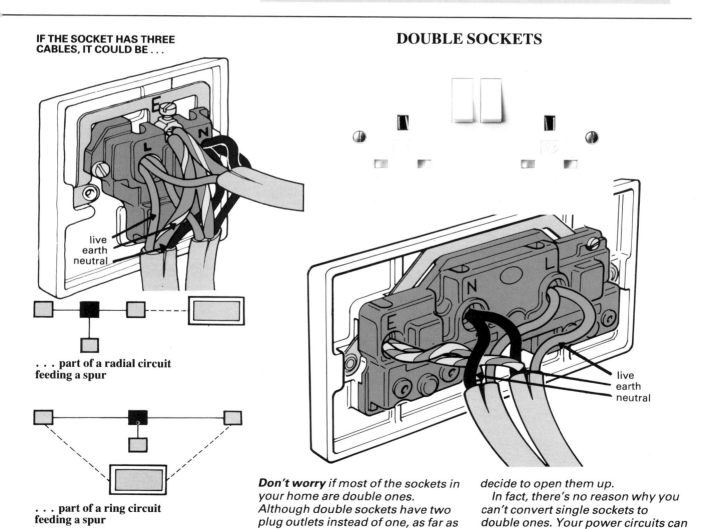

live
earth
neutral

. . . **part of a radial circuit feeding a spur**

. . . **part of a ring circuit feeding a spur**

DOUBLE SOCKETS

live
earth
neutral

Don't worry *if most of the sockets in your home are double ones. Although double sockets have two plug outlets instead of one, as far as the wiring goes they are identical to single sockets – as you'll see if you* decide to open them up.

In fact, there's no reason why you can't convert single sockets to double ones. Your power circuits can safely cope, and it's a lot safer than relying on adaptors.

17

FUSED CONNECTION UNITS

Fused connection units (FCUs) on power circuits are wired according to the same principles as ordinary sockets. It's quite likely that they are later additions to the main circuit, which explains why most FCUs are spurs.

However, because an FCU supplies an appliance direct (rather than through a plug and flex) there are important differences.

Some types of FCU supply the appliance via a length of circuit cable. This is likely to be hidden in the wall, but will be exposed if you open up the unit's faceplate (don't forget to switch off the mains first). The diagram below shows how to tell which is which.

Other types of FCU supply the appliance via a length of flex. This may or may not be hidden, but is easily identified by its different coloured cores (see diagram).

Fused connection units (right) come in various styles. Some are switched, some have an additional warning light, and others have a flex outlet on the front of the faceplate. All are fitted with their own plug-type fuse.

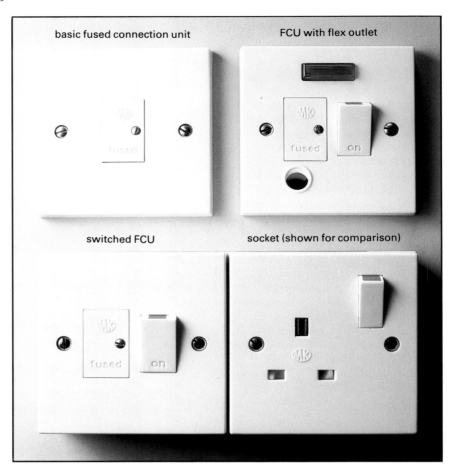

FCU WITH CABLE OUTLET

FCU WITH FLEX OUTLET

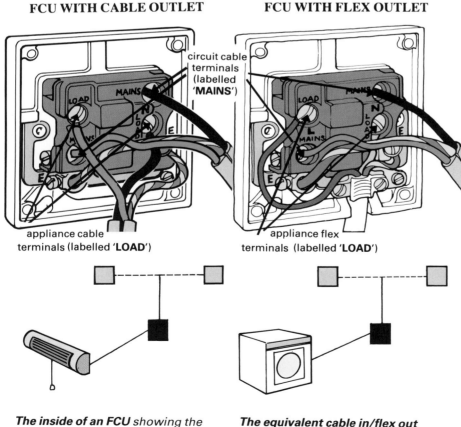

circuit cable terminals (labelled 'MAINS')

appliance cable terminals (labelled 'LOAD')

appliance flex terminals (labelled 'LOAD')

The inside of an FCU showing the wiring for a cable in/cable out arrangement. The two sets of terminals on the faceplate should be clearly marked.

The equivalent cable in/flex out arrangement. The flex may exit through or behind the faceplate, but is always held securely by a screw-down grip.

JUNCTION BOXES

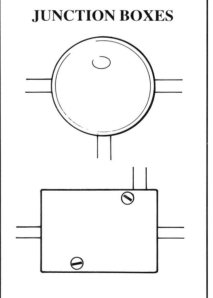

Junction boxes are used to make connections between circuit cables other than at a socket. **Round junction boxes** (top) are identical to those used for lighting circuits and are only found where the circuit wiring is hidden or buried. **Rectangular junction boxes** (above) are used where the circuit wiring is exposed.

TESTING ELECTRIC CIRCUITS

Testing electric circuits – either to find out if they are 'live', or simply to reassure yourself that they are safely wired – is one of the most important of all electrical jobs. The four occasions when you're most likely to do so are:

■ When you've just moved – to make sure the electrical system is safe.

■ When an appliance, light, socket outlet, or even a whole circuit, has stopped working.
■ When a fuse blows persistently.
■ When you're doing alterations and need to know if a particular circuit is live (or whether it's safe to wire into it).

Even if you don't plan to get involved in extensive electrical work, there's a lot you can tell simply by carrying out a visual check on the system (see below).

If you want to take things a stage further, you'll need at least some of the circuit testing devices described overleaf. These can be used to pinpoint a range of faults, from breaks in cables to poorly earthed electrical appliances.

MAKING VISUAL CHECKS

CONSUMER UNIT

Turn off the power at the main switch before checking for:

Main casing
☐ Cracks/exposed terminals.
☐ Non-fire resistant backing board (open-backed units).

Cables (in and out)
☐ Look for damage to the insulation. If you find any rubber/lead sheathed cable, see *Dangerous cables* below.

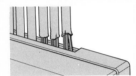

All fuseholders and MCBs
☐ Remove one at a time and examine for cracks or burning.

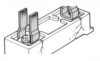

Rewirable fuses only
☐ Make sure the fuse wires are of the correct rating (compare with new fuse wire to check).
☐ Ensure the fuse wire securing screws are tight.

SWITCHES AND SOCKETS

Exterior
☐ Check faceplates for cracks. Look for charring around socket holes – a sign of overheating.

Interior
Check power is off, unscrew the faceplates and check for:
☐ Loose connections and bare cable cores at terminals.
☐ Missing green and yellow sleeving on earth cores.
☐ Missing earth connections between faceplate earth terminals and metal backing boxes (now required by the Wiring Regulations).
☐ Missing rubber grommets where cables enter a backing box (can cause chafing).
☐ Damaged or perished insulation (see *Dangerous cables* below).

LIGHT FITTINGS
Pendant lights
Check that the mains is off, unscrew the ceiling rose covers and look for:
☐ Loose wiring connections.
☐ Frayed flex sheathing and damaged cable insulation.
☐ Wiring not looped around cable anchors.
☐ Cracked or charred lampholders.

Other lights
Unscrew the battenholder or light fitting, or remove the shade and look for:
☐ Cracks and charring around the fitting/holder.
☐ A loose fit between the bulb and lampholder.
☐ Damaged cable insulation (cable exposed inside light fittings must be protected by heat-resistant sleeving).

CABLE RUNS
Inspect exposed cable runs in the loft, or beneath a suspended timber floor if there is space. Check for:
☐ Damage to cable insulation (including teeth marks left by rodents – a very real hazard).
☐ Rubber and lead-sheathed cable (see *Dangerous cables*).
☐ Flex used instead of cable.
☐ Wrongly sized cables.
☐ Cables not properly secured.

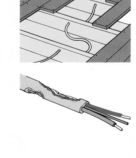

▲ **Dangerous cables**
Rubber and lead sheathed cable (left) becomes brittle with age and crumbles, leaving the cable dangerously uninsulated.
PVC sheathed cable is not immune to ageing either – the insulation hardens and breaks if exposed to heat. Cables in this condition must be renewed.

EQUIPMENT FOR TESTING CIRCUITS

PLUG-IN TESTERS

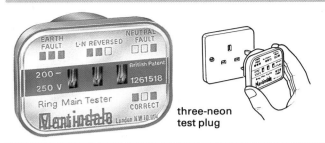

three-neon
test plug

Plug-in testers are simple devices which show whether a socket outlet is wired correctly – and if not, where the fault lies. They resemble three-pin plugs, but have two or – more usually – three neon lights set into the cover. When plugged into a live socket, various combinations of lit and unit neons give information on specific faults.

MAINS VOLTAGE TESTERS

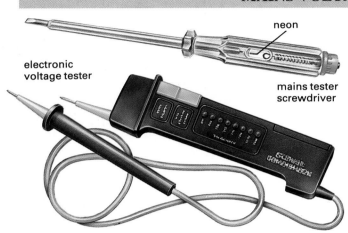

electronic
voltage tester

neon

mains tester
screwdriver

Voltage testers can tell whether a particular wire is dead, or has current passing through it. There are two types:
Mains tester screwdrivers look much like ordinary electrician's screwdrivers, but have a metal-capped hollow handle containing a neon bulb, a high-value resistor and a spring. When you touch the tip of the screwdriver to a live wire with your finger on the metal cap, a tiny current (reduced to safe levels by the resistor) flows up the blade and through your body to earth, lighting the bulb.
Electronic voltage testers come in several types. The simplest merely detect the presence of live components, but more sophisticated ones can be used as continuity testers too (see below).

Make sure that any tester you buy has most of the blade or probe insulated, or there is a risk you might accidentally touch live metal while testing.

CONTINUITY TESTERS

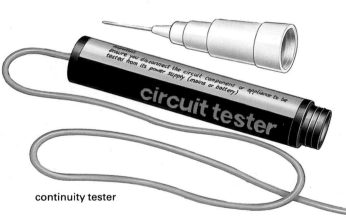

continuity tester

As their name suggests, continuity testers can tell you whether a conductor through a circuit, component, or single wire is continuous – or whether it is broken at some point due to a fault. They do this by passing a small current through the object concerned.

They look like an over-sized mains tester screwdriver: a wire with a crocodile clip on the end is connected to the handle, which contains a battery and bulb.

To use, simply connect the clip to one side of the circuit, and touch the tip of the screwdriver to the other side. If current can flow between the two points, the bulb will light.

DIY shops may not stock continuity testers, so try car accessory outlets, or electronics shops.

MULTIMETERS

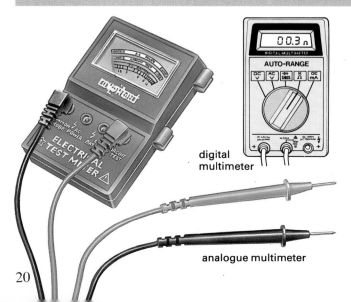

digital
multimeter

analogue multimeter

Essential to the electronics enthusiast, a multimeter can be used around the home as a sophisticated continuity tester – for checking appliance flexes, fuses and circuit cables – and also as a mains voltage tester.

The meter has two test leads with pointed probes on the ends which you touch to the component under test. The reading is shown on either an analogue swing meter or a digital display. Meters designed for household tests only have scales for these – eg 'continuity' or 'circuit test', low voltage (battery test), and 'mains live'. More expensive multimeters can also test other electrical values, such as current (in amps) and electrical resistance (in ohms).

If you feel these extra facilities make it a better buy than a simple continuity tester, choose a meter with a voltage range of up to 500V, and a resistance range of down to less than 1 ohm. For testing mains circuits you may need accessories – some manufacturers supply purpose-made extension leads and crocodile clips for very little extra.

USING A TEST PLUG

Using the test plug is simplicity itself. Just push it into the socket (switch on if the socket has an on/off switch) and watch the neon lights. The chart (see right) shows what the lights mean on one popular plug design – others may vary.

The only fault the plug can't detect is where the neutral and earth have been accidentally swapped over; to check this, you must turn off at the mains and open up the socket for a visual check. Check, too, for earth sleevings.

PLUG SHOWS	WHAT IT MEANS
▣ ☐ ☐	Socket is correctly wired up.
☐ ☐ ☐	The live wire is disconnected.
▣ ▣ ☐	Live and neutral connections are reversed.
▣ ▣ ▣	Either the earth connection is missing or live and earth are reversed.
☐ ☐ ▣	The neutral wire is disconnected.

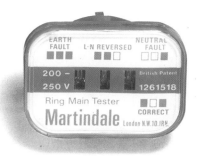

USING A MAINS TESTER

Use a mains tester to find out whether current is reaching a socket, light, switch or any other electrical item where you can touch the screwdriver tip to a live terminal. Before doing any tests, however, clean the tip on glasspaper and check the tester works by touching it to something known to be live.

To test a socket, make sure it is switched on, then open the safety shutters by inserting an insulated screwdriver into the earth (central top) pin hole.

Poke the blade of the tester into the bottom right hole with your finger on the cap, and the bulb should glow. If it doesn't, try the bottom left hole; if the bulb glows now, the live and neutral connections on the socket have been reversed; turn off the power and remove the faceplate to check.

To test other fittings, such as a light switch, turn off the power at the consumer unit and unscrew the faceplate. Check the wiring behind is secure, and that there is no possible risk of a short circuit, then turn the power back on again.

From now on, only hold the fitting by the edge of the faceplate. With your free hand, touch the tester blade on each terminal screw in turn.

The correct way to use a mains tester, with one finger on the metal cap and the rest well away from any live parts. Hold the disconnected faceplate by the edge only.

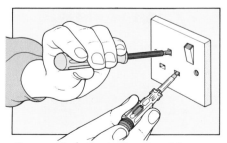

On a socket, the tester should only light in the bottom right hole. If it lights on the others, turn off the mains, remove the faceplate and check the wiring.

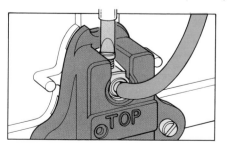

Check for current at a light switch by touching the tester to the recessed terminal screw. Take care not to strain the wiring as you twist the switch for access.

Trade tip

Good contacts

❛ With all testers, especially blunt-tipped mains tester screwdrivers, make sure you get a good contact with the part being tested. Dirt or corrosion can easily give a false reading, even if the metal looks 'bare'. The probes on other testers tend to be more reliable, but in this case the low voltages used can cause similar problems. ❜

USING A CONTINUITY TESTER

A continuity tester allows you to check whether a fuse, a length of flex or cable, or virtually any electrical component, is capable of conducting electricity. But remember, this isn't a foolproof test of the item's ability to handle mains voltages (see safety panel).

When testing circuits or appliances, work out a systematic testing sequence so that you can narrow down the fault to a particular section or component.

To test circuit cables, start by turning off the power at the consumer unit. Then unscrew two of the socket or switch faceplates on the faulty circuit.

Clip the tester lead to the cable live core at one of the fittings (use an extension lead if necessary) then probe the live core at the other. If the connection between the two is sound, the tester will light.

To test appliances, connect the tester across the flex terminals, motor, element or any other accessible parts. No light at any stage means that a component or connection within the part of the circuit being tested has failed.

You can also test the appliance wiring and earth connections.
- With the appliance unplugged, test the fuse hasn't blown.
- Connect between the earth pin of the plug and the appliance's metal casing. If the light glows, the earthing is satisfactory.
- Now connect between the plug's live pin and the casing; a glow here indicates faulty insulation.

Mains cable continuity Connect the continuity tester to each end of the same cable core (here it's the live) where the cable emerges from the wall.

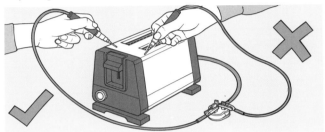

Metal-cased appliance Test the earth between the plug's earth pin and the metal casing. Check the live insulation between the live pin and casing.

USING A MULTIMETER

Multimeters can be used instead of continuity testers, and they can also be used to check whether batteries have any life in them, or if low-voltage circuits are live.

Multimeters are 'polarity-conscious' which means that when testing voltages the probes must be connected the right way round – always connect the red probe to the live (+) terminal and the black probe to the neutral (−).

Before testing anything, make sure the function dial is set correctly. For checking battery voltage select 'low volts'; to check continuity set it to 'ohms' or 'circuit test'.

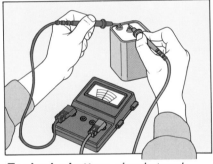

To check a battery, simply touch the meter probes to the terminals after setting the function dial to 'low volts'. The voltage will be displayed on the meter.

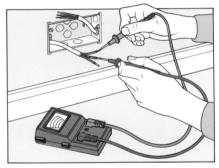

Test for cable short-circuits by touching one probe to the live core and the other to earth and neutral in turn. There is a fault if the meter reading drops.

To check continuity, connect the probes to each end of the circuit. The reading will drop to near zero (ohms) or move into the 'yes' sector.

Circuit tester drawbacks

The drawback of continuity testers and multimeters is that they work on a low voltage. This means that they won't show up faults in the wiring such as bare cable cores which are very close to each other, but not touching. At mains voltage the current could arc across the gap, causing the wiring to overheat and even start a fire.

So while you can rely on testers to show up major defects – live and earth cores touching, or broken cores, for example – don't automatically assume a circuit or appliance is 100 per cent safe without further investigation. In many cases, a visual check is more reliable.

Professional checking

Don't hesitate to call in an electrician if you are unsure what has caused a fault. The professionals use high-voltage testers (see left) that can check a circuit for complete continuity and insulation resistance, as required by the Wiring Regs.

Qualified electricians also have equipment for testing the effectiveness of the earth circuit, and the operation of residual current circuit breakers fitted to the system.

Where a house has been fully rewired, the system is tested as a matter of course by the Electricity Board's engineers. They will only connect to the mains once they are satisfied everything is safe.

CABLE AND ACCESSORIES

Household wiring is divided into two types. **Cable** is used for the fixed wiring running from the consumer unit to power circuits, lighting circuits and special circuits for high-powered appliances. Don't confuse it with **flex** (short for flexible cord), which is used to connect portable appliances and table lamps to plugs, and for the final connections to fixed appliances and lights.

Cable contains individual copper conductors called *cores*, all of which apart from the earth core are insulated with PVC sleeving. The earth core is left bare, except at junction boxes, socket outlets and light fittings etc, where it is covered with slip-on green and yellow PVC sleeving (sold separately).

Types of cable

The most commonly used type of cable – **PVC sheathed** – has the cores enclosed in a tough layer of PVC. There are two versions:
Two-core and earth cable has insulated live and neutral cores, plus a bare earth core. It is used for both power and lighting circuits.
Three-core and earth cable has a third insulated core for multi-way switching on lighting circuits only.
Armoured cable is designed for laying underground and has a much tougher outer casing to protect it from damage. However, it is expensive, and requires special fittings at connection points. The alternative is to run ordinary PVC-sheathed cable enclosed in conduit.

Cable sizes

All types of cable are available in various sizes, measured by the cross-sectional area of their cores in

square millimetres (mm^2). This figure is either printed on the cable sheath itself, or on the reel.

The thickness of the cores is the main factor determining how much current a cable can carry safely, which is why cable for a circuit rated at 30 amps must be considerably thicker than cable for a circuit rated at 5 amps. But outside factors can also affect a cable's safe current carrying capacity (see below), so treat the sizes given for different fuse ratings on the chart overleaf with caution – and never use a size smaller than the one specified.

Be sure, too, not to confuse the cable's safe capacity with its *maximum current rating* – the amount of current needed to melt it!

Small quantities of cable are usually bought 'off the reel' by the metre length.

How safe is safe?

Because cable heats up as current passes through it, a cable which is enclosed in some way may have a lower current carrying capacity than an exposed one of the same size. In plaster, the difference is normally small, but if the cable runs through insulation you may need a larger size. The length of a cable can also affect its capacity, which is why some sizes in the chart relate to the area of a circuit – not its fuse rating.

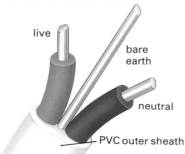

Cable cores are single- or seven-stranded depending on the size. The live and neutral cores are insulated, but not the earth. Don't confuse cable with . . .

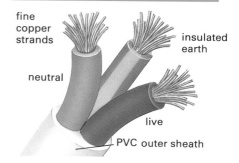

. . . *flex*, which has insulated cores made up of many thin strands of wire. The round-section outer covering is flexible enough to bend.

PVC SHEATHED CABLE

Twin-core and earth cable has a red insulated live core and a black insulated neutral core, plus a bare earth core.

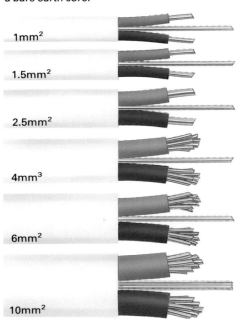

1mm²

1.5mm²

2.5mm²

4mm³

6mm²

10mm²

Three-core and earth cable has a red insulated live core, plus blue and yellow insulated switch cores.

1mm²

CABLE SIZE	FUSE CURRENT RATING (rewirable fuse)	(cartridge fuse or MCB)	USED FOR
1mm²	5 amp	5/6* amp	■ Lighting circuits carrying not more than 1.2kW
	—	10 amp	■ Lighting circuits carrying not more than 2.4kW
1.5mm²	—	10 amp	■ Lighting circuits carrying not more than 2.4kW
2.5mm²	15 amp	15/16* amp	■ Fused spurs
	15 amp	15/16* amp	■ Immersion heater
	—	20 amp	■ Radial final (main) circuits serving up to 20 sq m (22 sq yd) of floor space
	30 amp	30/32* amp	■ Ring final (main) circuits serving up to 100 sq m (111 sq yd) of floor space
4mm²	—	30/32* amp	■ Radial final (main) circuits serving up to 50 sq m (55 sq yd) of floor space
		30/32* amp	■ Ring final (main) circuits serving up to 100 sq m (111 sq yd) of floor space
			■ Showers up to 7kW
6mm²	30 amp	45 amp	■ Cookers up to 12kW
	—	45 amp	■ Showers up to 10.5kW
10mm²	40 amp	45 amp	■ Cookers over 12kW
1mm²	—	5 amp	■ Multi-way switch part of lighting circuit

*new sizes now available

NB *Cables are shown actual size. Sizes given are absolute minimums – ask an electrician if in doubt what size to use.*

ARMOURED CABLE

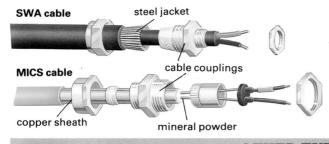

SWA cable — steel jacket, cable couplings

MICS cable — copper sheath, mineral powder

Domestic armoured cable comes in 6mm² and 10mm² sizes, and in two- or three-core versions. There are two types:
Steel wire armoured (SWA) cable has a protective steel jacket around the insulated cores. The whole lot is encased in a thick weatherproof PVC sheath, and there are special couplings for connecting the cable at junction boxes.
Mineral insulated copper sheathed (MICS) cable has two cores insulated with mineral powder. These are then encased in a PVC-covered copper sheath which forms the earth conductor. Connecting MICS cable is a professional job.

OTHER TYPES OF CABLE

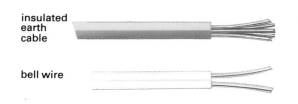

insulated earth cable

bell wire

Insulated earth cable has a single multi-strand core enclosed in a green/yellow PVC sleeve. There are three sizes: 16mm² earthing conductor for connecting main earth terminal to electricity board's earth; 10mm² for cross-bonding between the main earth terminal and gas or water pipes; and 6mm² for earthing the consumer unit to the main earth terminal.
Bell wire is for wiring doorbells to an extra low-voltage supply from a transformer. Do not use for mains wiring.

CABLE ACCESSORIES

junction box

earth insulating sleeve

plastic clip

self-adhesive clips

Green and yellow sleeving is slipped over the bare earth cores at light fittings, junction boxes and sockets etc to guard against short circuits. The size varies, but isn't critical.
Cable clips are used to secure surface-mounted wiring.
Plastic clips have hardened steel pins that hold securely in wood, plaster and render. They are ideal for most jobs, and come in sizes matched to the various cable sizes.
Self-adhesive clips are available for places where you can't hammer pins into the wall.
Junction boxes are for joining cable. All the joints – including the earth conductor – must be inside the box.

CHOOSING FLEX

Flex (flexible cord) is used to connect all movable appliances to the fixed wiring, other than an electric cooker which is wired with cable instead. It is also used for table and standard lamps, pendant lights and chandeliers.

Flex is made up of several multi-stranded copper *cores* inside insulated sleeving. The insulated cores are then encased in a PVC or rubber outer sheath for protection. Most flex is either *two-core* or *three-core*, but *four-* and *five-core* is used in central heating systems.

Which flex to use?

Choosing the right flex depends on several things:

■ Whether the appliance or light needs an earth. Older appliances or metal-cased lights must have a three-core flex with an earth core. Many modern domestic appliances are *double-insulated*, which means they have no exposed metal parts. They don't require an earth and are wired with two-core flex.

■ The wattage of the appliance (and hence the current it draws). This determines the size of the flex, measured in square millimetres (mm^2). Each size has a maximum current rating, although in practice you should leave some margin for safety (see tables overleaf).

■ How hot the appliance gets. The vast majority of appliances and table lights can be safely wired with

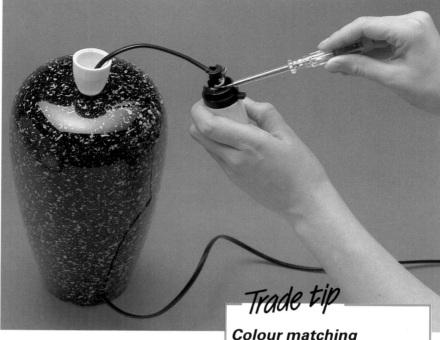

PVC insulated and sheathed flex which is heat-rated at 60°C. Pendant lights and modern irons are wired with *heat-resistant PVC insulated and sheathed flex* which looks the same but is designed to cope with hotter conditions (up to 85°C).

Older electric fires and some irons are wired with a special *heat-resistant braided flex*. A new flex should be the same type.

■ The weight of the fitting (pendant lights only). A heavy light needs thicker flex.

Trade tip

Colour matching

❛ Most PVC sheathed flex is coloured white, and while this suits many modern appliances it often clashes with older ones or sticks out from the room's colour scheme. However, other colours such as black, pink and gold are available.

You may have to scour the local electrical shops to find exactly what you want, but even if they don't hold it in stock, many good suppliers will order a special flex if you ask. ❜

REPLACING OLD FLEX

The flexes on old light fittings or appliances may have the old type of colour-coding – or even no coding at all. You may also find flex which no longer satisfies current standards. In all cases, the flex should be replaced with a modern equivalent.

PVC covered figure-of-eight flex resembles bell wire and has no outer sheath. This was often used on table lamps, but no longer meets the British Standard. Replace it.

Fabric covered twin-core twisted flex was used on many light fittings but the fabric is prone to fray and the insulation may have perished. If you find any flex like this, replace it as soon as

possible; although fabric covered flex is still available, a PVC sheathed flex is more robust.

Neither of the above has colour-coded cores. Replace with modern two-core flex (or three-core if the fitting is metal). Wire the plug or rose following the new coding – brown to live, blue to neutral (and green/yellow to earth). Lampholder connections go either way round.

The old UK system of colour-coding used a red core for live, black for neutral and green for earth. If you find flex like this, rewire the appliance with a modern flex of the correct rating, using brown for live, blue for neutral, green/yellow for earth.

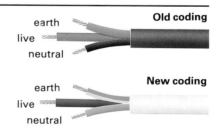

Old and new colour coding for three-core flex. Red becomes brown, black becomes blue and green becomes yellow/green.

fabric-covered twin-core twisted flex

PVC insulated twin-core 'figure-of-eight' flex

PVC INSULATED AND SHEATHED FLEX

Two-core flex has a brown insulated live core and a blue insulated neutral core. The outer sheathing may be round or oval in shape. When wiring pendant lights, ensure *heat-resistant* flex is used (this is visually identical).

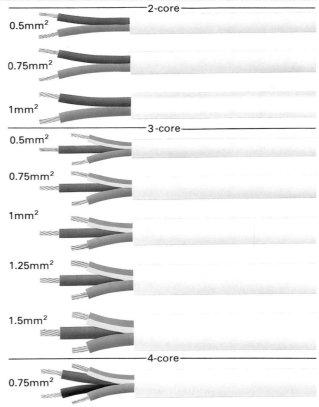

SIZE	AMPS	WATTS	TYPICAL USE
0.5mm²	3A	720W	Appliances up to 500W. Non-metal lamps with flex up to 2m (7') long. Lightweight pendants.
0.75mm²	6A	1.4kW	Appliances up to 1kW. Non-metal lamps with flex over 2m (7') long. Medium weight pendants.
1mm²	10A	2.4kW	Appliances up to 2kW. Heavyweight pendants.

Three-core flex has a brown insulated live core, blue insulated neutral core, plus a green/yellow insulated earth core. The outer sheath is round. When wiring pendant lights, irons, and storage and immersion heaters, ensure *heat-resistant* flex is used (this is visually identical).

SIZE	AMPS	WATTS	TYPICAL USE
0.5mm²	3A	720W	Metal cased lights and appliances up to 500W.
0.75mm²	6A	1.4kW	Metal cased lights and appliances up to 1kW.
1mm²	10A	2.4kW	Kettles and other appliances up to 2kW.
1.25mm²	13A	3.1kW	Heavy duty extension leads, kettles and other appliances up to 3kW.
1.5mm²	15A	3.6kW	Extra heavy duty extension leads, storage heaters and immersion heaters.

Four and five-core PVC sheathed flex is used for central heating and control circuits (though *tough rubber sheathed flex* – see below – may be used instead). The controls themselves – valves, thermostats etc – may employ a variety of flexes with unusual colour-codings, but only four-core is readily available. This comes in the 0.75mm² size, for currents up to 6A (1.4kW).

HEAT-RESISTANT BRAIDED FLEX

Heat-resistant braided flex is used on irons, electric fires, and any other appliances that might become hot and burn or melt a normal PVC sheathed flex. Available only in three-core, the outer sheath is made of heat-resistant rubber or PVC and is surrounded by a braided textile cover.

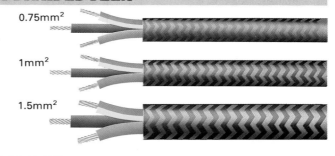

SIZE	AMPS	WATTS	TYPICAL USE
0.75mm²	6A	1.4kW	Appliances up to 1kW.
1mm²	10A	2.4kW	Appliances up to 2kW.
1.25mm²	13A	3.1kW	Portable fires up to 3kW.
1.5mm²	15A	3.6kW	Portable fires up to 3kW.

TOUGH RUBBER-SHEATHED FLEX

This has a thick black rubber outer sheath and rubber insulated cores. It is the standard flex for power tools or extension leads, where there is a risk of the lead becoming damaged. You may also find four- and five-core versions in central heating control circuits, in the 0.75mm² size.

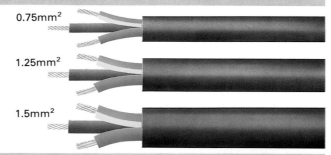

SIZE	AMPS	WATTS	TYPICAL USE
0.75mm²	6A	1.4kW	Tools and extension leads up to 1kW.
1.25mm²	10A	2.4kW	Tools and extension leads up to 2kW.
1.5mm²	15A	3.6kW	Tools and extension leads up to 3kW.

COILED LEADS

Coiled leads are a form of PVC sheathed flex, made in both two- and three-core versions. They are normally sold by current rating rather than size. The lowest current rating is 1A (used for shavers), ranging up to 6, 10 and 13A. Applications for the larger sizes are the same as for standard PVC sheathed flex.

Available in bright colours as well as traditional white, coiled flex is often used for its cosmetic value on table lights and rise and fall pendant lights. But it has a practical value too: when fitted to worktop appliances such as a kettle, its natural 'spring' stops the lead trailing over the edge.

10A coiled flex

PLUG AND FLEX REPAIRS

Plugs and flexes get a lot of heavy use. Although they may have been fitted correctly, damage and wear can make plugs unsafe, while those on older appliances may have been wrongly fitted in the first place. The picture below shows some of the hazards which may be lurking unsuspected, all of which can be put right for very little time and money.

Most appliances are connected to the mains via a plug and three-core (earthed) flex. The exceptions are double insulated appliances which don't need an earth, and lamps with no metal parts. Both of these may be wired with two-core flex. Lamps with metal parts should always be earthed.

If an appliance is rarely or never unplugged (eg a waste disposer or tumble drier), the alternative is to wire it directly to the mains via a fused connection unit. This is more reliable, and avoids tying up a socket permanently.

There are several kinds of flex for connecting different appliances and lights. Don't confuse them with *cable*, which is used for fixed wiring behind walls and under floors.

All flex has fine stranded wire conductors, and the insulated cores are coloured differently from cable.

Flexes and colour coding

Modern three-core flex has two cores coloured BROWN for LIVE and BLUE for NEUTRAL, plus a GREEN/YELLOW striped core which is the EARTH conductor. Two-core flex has no earth conductor. The cores are either coloured brown and blue, or left uncoloured.

Old three-core flex has a red Live core and a black Neutral core (the same as cable), plus a plain green Earth core. Flex old enough to have this colouring should be checked to make sure it has not deteriorated.

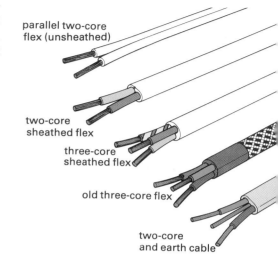

parallel two-core
flex (unsheathed)

two-core
sheathed flex

three-core
sheathed flex

old three-core flex

two-core
and earth cable

Modern flex colour codes are distinctive. Red and black insulation is used on old flexes and on modern cable. Uncoded flex is mainly used for lighting.

PLUG AND FLEX FAULTS

Wrong fuse fitted in plug: if too high-rated may not blow quickly if there is a fault. Change fuse.

Badly fitted plug: flex not gripped securely and wires are loose. Could pull out leaving live wire exposed. Rewire.

Extension cable overloaded: insulation could melt. Use extensions safely or rewire to avoid using them at all.

Cracked plug: terminals could become exposed. Replace with new one, possibly tough rubber type to withstand knocks.

Flex too long: can be accidentally pulled down by children, trailed across hob or dropped in sink. Shorten flex.

Damaged flex: insulating tape repairs are potentially dangerous. Replace flex or join with a flex connector.

FITTING PLUGS AND FUSES

Nowadays, appliances are normally connected using identical plugs with three square pins – but the fuse in the plug is changed to suit its purpose (see opposite).

Old type round-pin plugs were unfused and their presence generally indicates a wiring system which is old enough to need renewing.

In the old system there were three different sizes of plugs – 2 amp for lighting, 5 amp for low current appliances (up to 1kW) and 15 amp for high current (1-3kW) appliances.

Plastic plug: Normally the cheapest type. The pins are often plain brass and the flex grip is usually a strap tightened with two screws.

Safety-type plug: Flex grip is built in and the pins are partially insulated to prevent accidents caused by touching live parts when pulling out the plug.

Unbreakable plug: Has a durable casing made of tough rubber to resist knocks and heavy use. Particularly good for things like power tools.

Easy-fit plug: Various patterns. All wires are normally the same length. Terminals and cord grip are designed to work without needing a screwdriver or stripper.

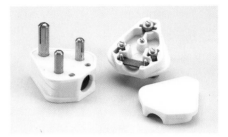

Round-pin plugs: Unfused, made in different sizes for different ratings. Only used with old installations; probably indicates rewiring is needed soon.

WIRING A PLUG – THE RIGHT WAY

Fitting plugs safely is a matter of taking care to get the wires the right length, and using the right tools to fit them. You probably need to cut and strip the wires even when the appliance manufacturers have stripped them for you. Because plugs differ so much, it's very unlikely that the ends will be exactly the length you need to do the job properly, and in many cases you are better off cutting them short and starting again.

The correct lengths for the wires depend on your plug. On several types of plug, all the wires are trimmed to the same length. But on most plugs the earth wire is longer, and the live and neutral may also be different from each other.

There may be instructions on the packaging to tell you the correct lengths of inner and outer insulation to strip. Otherwise try the wires in place to see that they fit neatly into the flex grip and plug channels as shown below.

CORRECT PLUG WIRING

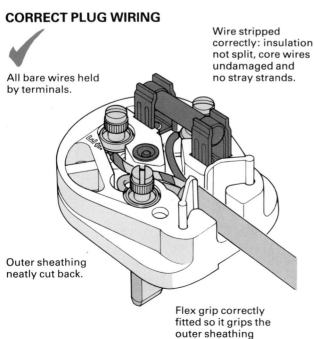

All bare wires held by terminals.

Wire stripped correctly: insulation not split, core wires undamaged and no stray strands.

Outer sheathing neatly cut back.

Flex grip correctly fitted so it grips the outer sheathing firmly.

UNSAFE PLUG WIRING

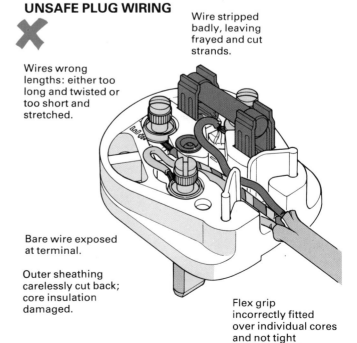

Wire stripped badly, leaving frayed and cut strands.

Wires wrong lengths: either too long and twisted or too short and stretched.

Bare wire exposed at terminal.

Outer sheathing carelessly cut back; core insulation damaged.

Flex grip incorrectly fitted over individual cores and not tight enough to grip.

USE THE RIGHT FUSE

It's important to fit a fuse with the correct rating. The fuse protects the wiring from overloading if there is a fault or short circuit. It is designed to form a 'weak link in the chain' and blow before the flex or internal wiring in the appliance itself can overheat. If you fit one designed for a higher current than is needed, there is a risk that it may not blow if there's a fault.

Plug fuses are normally rated at either 3 amps or 13 amps. 3 amp fuses are coloured red and are used for low-powered appliances drawing up to 720 Watts (720W). 13 amp fuses are coloured brown and are used for any appliances drawing from 720W to 3000W (3kW). Some examples of low and high power apparatus are shown below, but you should check the rating plate on the appliance (see right).

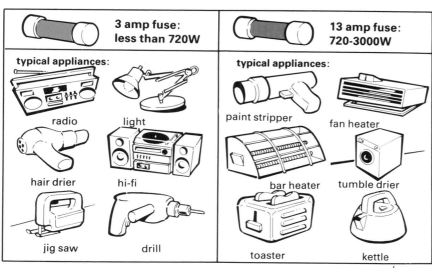

3 amp fuse: less than 720W

typical appliances: radio, light, hair drier, hi-fi, jig saw, drill

13 amp fuse: 720-3000W

typical appliances: paint stripper, fan heater, bar heater, tumble drier, toaster, kettle

HOW TO READ THE RATING PLATE

Somewhere on the appliance should be a rating plate giving useful information. The most important is the power rating (measured in Watts), which governs the type of fuse (and flex) needed. It should say something like 'W 350' (350 Watts) or '2000W at 240V' (2000 Watts). Some high ratings may be in kilowatts (1kW=1000 Watts) and low power apparatus (like a radio adaptor) may give the current in mA (milliamps) instead. Fit a 3 amp fuse to anything this low powered.

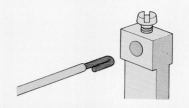

voltage and frequency requirements — approvals symbol

ELECTRI Co
Model 203
240V~50 Hz
3000 Watts
Made in UK

power or current rating in Watts or kilowatts — double insulated symbol (appliance needs no earth)

STRIPPING/CONNECTING

You need an electrical screwdriver, trimming knife, and something to strip the wires. It's possible to manage with a knife alone, but there's a risk of damaging the wires accidentally. A cheap pair of wire strippers makes the job much easier and helps to ensure all your wiring is as safe as possible.

Cut back any outer sheathing leaving the inner cores long enough to reach the terminals and make the connections. The right amount of the core insulation to strip depends on the terminals (right).

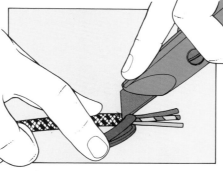

1 Use a knife to slice the insulating sheath and peel it open. If there's a woven cover pull this away. Bend the sheath back and cut away the surplus.

2 If the flex has a woven cover, wind on insulating tape to prevent fraying or roll back the inner rubber sheathing over the woven sheath for a short way.

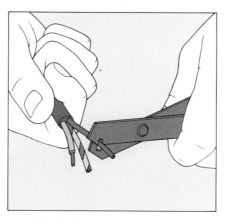

3 Cut the inner cores with wire strippers. Strip off the right amount of insulation, ensuring you don't cut the wire. Twist the strands to prevent fraying.

How you connect the wires depends on the terminals.

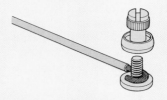

Post type: bend the bared end back on itself then push into the hole. This prevents the ends of the wire snagging on the hole.

Screw type: wind the end of the wire clockwise around the screw. Make sure the end of the flex is trapped by the clamp nut.

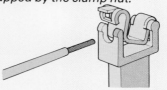

Clip type: push the bare end into the clip. Some patterns have a snap-on catch, others lock when the cover is replaced.

EXTENDING FLEX

Extension leads are intended for temporary use. If you need a long flex on one particular appliance for most of the time, then ideally you should have the existing flex replaced with a longer length of the same type or avoid the problem by adding a new socket. But if this isn't possible, you can extend the existing flex using a *flex connector* or an *in-line switch*.

A flex connector can also be used to join a flex which has been accidentally damaged, after first cutting out the damaged section. However, it's preferable to wire the appliance with a new flex if you can.

Flex connectors come in two types – fixed and plug-in. Both normally have three terminals, although there are two-terminal versions for double insulated appliances and lamps with no metal parts.

In-line switches are a sensible alternative for appliances such as bedside lights, since they both extend the cable and put control of the light within easy reach.

With either type of extension, make sure the new length of flex is as near identical to the old one as possible. Flex differs in its current rating and in the durability of its insulation, both of which vary to suit the application.

A coiled extension lead is useful as a temporary flex extender, but should not be used all the time – especially for appliances which take a lot of power.

A multi-socket block makes a more versatile extension, but will be overloaded if the total power drawn by the appliances plugged into it exceeds 3000 Watts (3kW).

WIRING A FLEX CONNECTOR OR IN-LINE SWITCH

Check the current rating of the connector you buy is suitable for your appliance (see page 29). Fixed connectors are normally rated at 13 amps, 3000 Watts (3kW). Plug-in types may be rated at a maximum of 5 Amps, 1200 Watts (1.2kW).

Follow any labelling on the connector, or instructions stating which way round it should be connected. Terminals may be labelled L, E, and N for Live, Earth and Neutral. Plug-in types should state which side is connected to the mains. This is essential for safety.

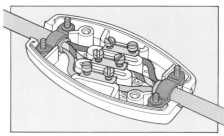

1 **To fit a fixed flex connector** unscrew the cover and loosen the screws of the two flex grips at either end. Check the length of wire you need to strip.

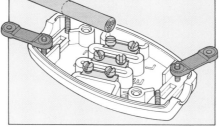

2 Strip the ends of both lengths of flex and insert them into the connector under the flex grips. Tighten these later to clamp the outer sheath.

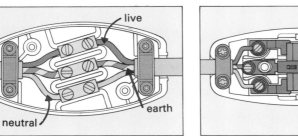

3 *The terminals may not be labelled. Connect the earth wires to the centre terminals and connect brown to brown and blue to blue on the outer terminals.*

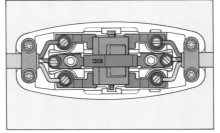

An in-line switch is fitted in almost the same way as the fixed flex connector, but the terminals are connected to the switch itself rather than to each other.

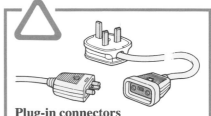

Plug-in connectors

These must be wired so that the plug part (with projecting pins) connects to the appliance. If you connect this half to the mains, the pins could become live when the connector is pulled apart.

REPAIRING BLOWN FUSES

Home electrical systems have a series of weak links built into them to protect the wiring from overheating in the event of an overload or short circuit.

■ Virtually every plug-in appliance has its own fuse inside the plug, and there are similar fuses inside fused connection units (FCUs) and other devices. These will protect the circuit from appliance faults and vice versa.

■ Inside the consumer unit (fuse box) are the fuses or miniature circuit breakers (MCBs) which protect the main lighting and power circuits.

Tracing fuse faults

Repairing or replacing a blown fuse is simple enough, but for safety's sake it's essential to find out what caused it to blow in the first place.

Use the chart below to help you. The fault is more likely to be in the appliance than the circuit, so check the plug fuse first (see page 34). Sometimes, however, appliance faults cause the circuit fuse to blow too.

You may need an electrician to confirm your suspicions by testing the relevant circuits. Even so, you'll save time and money if you can point them in the right direction.

Circuit fuse faults

There are two faults which can be cured simply by repairing the circuit fuse (or resetting the MCB):

Power surges occur naturally, blowing fuses that have become oversensitive with age.

Overloads mainly affect old radial circuits which were not designed with modern power-hungry appliances in mind. The maximum power rating of a radial circuit is 7200W (7.2kW), so if you suspect an overload, add up the wattages of all the appliances on the circuit (given on their rating plates) and check that the total does not exceed this figure.

Taking precautions

Fuses can blow at any time, so don't be caught unprepared:

■ Get to know your consumer unit: check what type of fuses are fitted (see overleaf), then buy spares for each rating or a card of fuse wire as appropriate. If you haven't already done so, label each fuse or MCB with the circuit it protects.

■ Buy packs of spare plug fuses.

■ Store the spares, together with a torch and a screwdriver, near the consumer unit.

TRACING FUSE FAULTS

To use the chart, answer the question on the right. Then follow the green arrow for a 'Yes' answer, or the red arrow for a 'No' answer. Continue in this way, answering questions and following the appropriate arrows, until you've traced the fault.

NB Some systems have *Residual Current Devices* (RCDs) as well as fuses (see overleaf). These don't affect the chart.

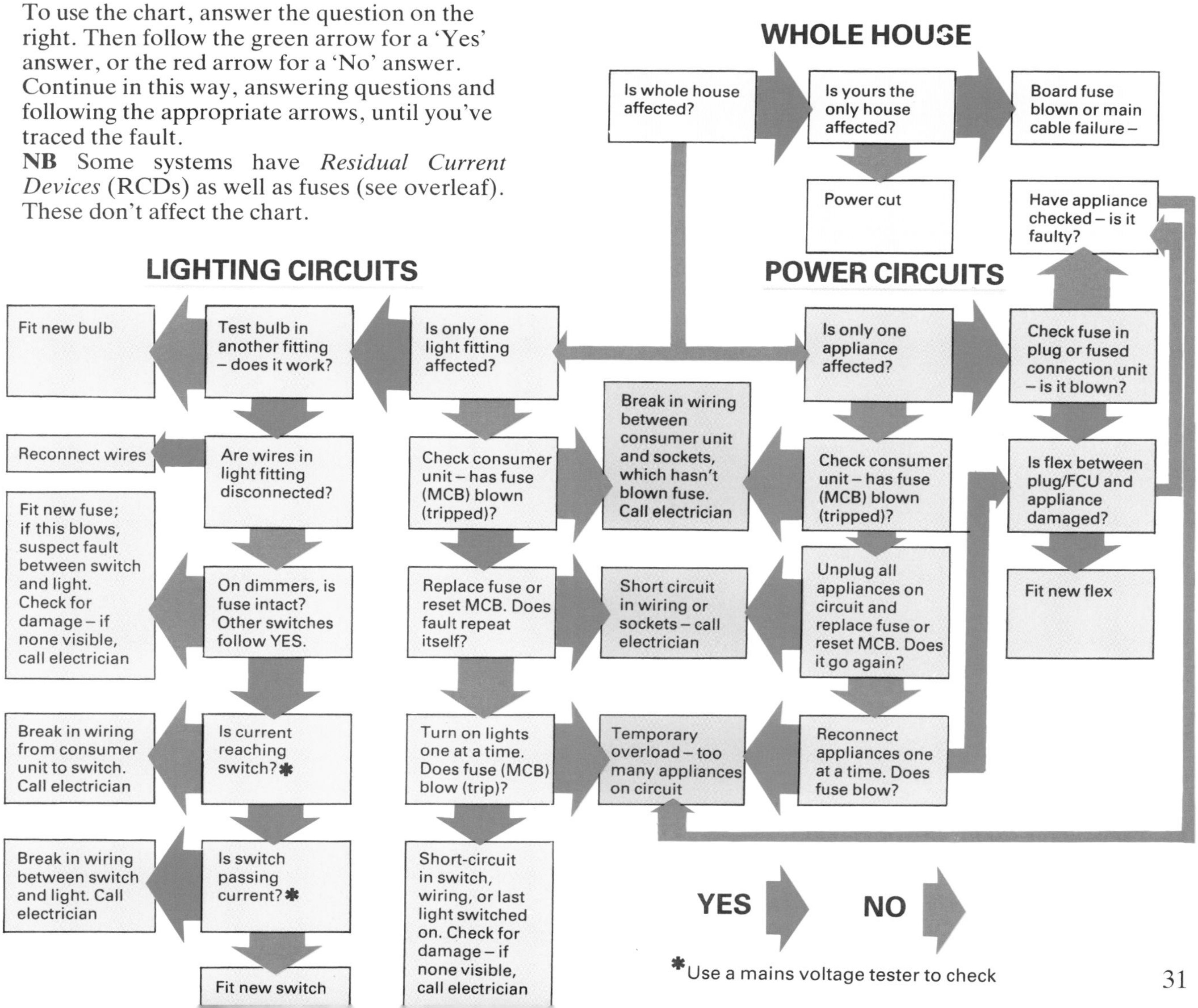

WHOLE HOUSE

Is whole house affected? → Is yours the only house affected? → Board fuse blown or main cable failure –

Power cut

Have appliance checked – is it faulty?

LIGHTING CIRCUITS

Fit new bulb ← Test bulb in another fitting – does it work? ← Is only one light fitting affected?

Reconnect wires ← Are wires in light fitting disconnected?

Fit new fuse; if this blows, suspect fault between switch and light. Check for damage – if none visible, call electrician

Check consumer unit – has fuse (MCB) blown (tripped)?

On dimmers, is fuse intact? Other switches follow YES.

Replace fuse or reset MCB. Does fault repeat itself?

Break in wiring from consumer unit to switch. Call electrician

Is current reaching switch?*

Turn on lights one at a time. Does fuse (MCB) blow (trip)?

Break in wiring between switch and light. Call electrician

Is switch passing current?*

Short-circuit in switch, wiring, or last light switched on. Check for damage – if none visible, call electrician

Fit new switch

Break in wiring between consumer unit and sockets, which hasn't blown fuse. Call electrician

Short circuit in wiring or sockets – call electrician

Temporary overload – too many appliances on circuit

POWER CIRCUITS

Is only one appliance affected? → Check fuse in plug or fused connection unit – is it blown?

Check consumer unit – has fuse (MCB) blown (tripped)?

Is flex between plug/FCU and appliance damaged?

Unplug all appliances on circuit and replace fuse or reset MCB. Does it go again?

Fit new flex

Reconnect appliances one at a time. Does fuse blow?

YES ▶   NO ▶

*Use a mains voltage tester to check

31

REPAIRING CIRCUIT FUSES

The consumer unit contains a bank of circuit fuses or MCBs mounted beneath a protective cover. On modern units the cover simply hinges out of the way, but in some older models it is held by screws.

The fuses or MCBs have different ratings, depending on whether they protect lighting circuits, power circuits, or special circuits for individual high-power appliances.

Always replace a fuse with one of the same rating (see chart on page 34). And whatever the fuse, always turn off at the main switch before removing or replacing the holder.

Fuses and circuit breakers

When there is an electrical fault in a fuse-protected circuit, it creates a surge of current which eventually causes the fuse to melt – or trip, in the case of an MCB. But this doesn't always happen immediately: it's still possible for the wiring to be damaged – or for a person to be electrocuted – before the surge grows large enough to blow the fuse.

For this reason, many newer consumer units have the added protection of a resettable *Residual Current Device* (RCD). RCDs detect the leakage of current to earth which occurs whenever there is a fault in the circuit – at which point they trip, cutting off the power. They react much faster than fuses, and are therefore safer.

Kits are available for converting older systems to RCD protection, and you can buy plug-in versions for individual appliances.

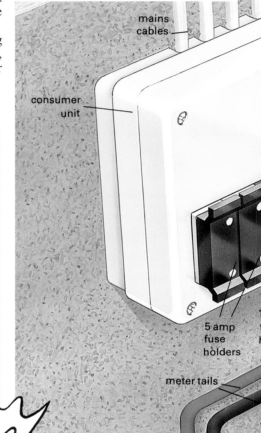

mains cables

consumer unit

5 amp fuse holders

meter tails

electricity meter

No bodging

Never try to repair a blown fuse by fitting a fuse or fuse wire of a higher rating – even in an emergency. Likewise, never use silver paper, paper clips or nails to take the place of a fuse or fuse wire. Such objects may not melt in the event of a fault – with potentially disastrous results.

REWIRABLE FUSES

There are three main designs of rewirable fuse, but they all work in the same way. A wire link (the *fuse wire*) bridges the two terminals of the ceramic or plastic holder, and melts if a fault occurs.

Turn off the power at the main switch and remove the consumer unit cover. Pull out the fuse holder and see whether the fuse wire has burnt – on the enclosed type there is a viewing window in the porcelain tube.

Replacing the wire

■ The fuse wire is secured by a screw at each end of the holder. Undo the screws completely (watch out for the small metal washers fitted beneath the screw heads) and pull out the burnt ends.

■ Cut a length of fuse wire of the correct rating (see chart overleaf).

■ Feed the wire between the two terminals. This is slightly fiddly on the enclosed type, since the wire needs to be fed through holes in the holder.

■ Refit the screws loosely and twist the ends of the fuse wire around them in a clockwise loop.

■ Tighten the screws just enough to hold the fuse wire and trim off any excess wire.

■ Replace the holder in the consumer unit, refit the cover, then turn on the power supply.

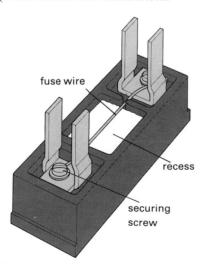

fuse wire

recess

securing screw

In exposed fuse holders, the fuse wire simply runs across a recess between the two terminals.

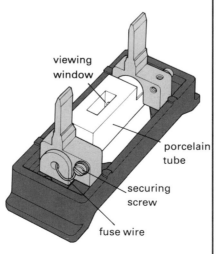

viewing window

porcelain tube

securing screw

fuse wire

In protected fuse holders, the fuse wire runs through a porcelain tube.

Trade tip

Leave some slack

6 When you fit new fuse wire to a rewirable fuse don't pull the wire too tight. On thin, 5 amp fuse wire in particular, there is a risk that it will stretch and snap as you tighten the screws. Leave a little slack. 9

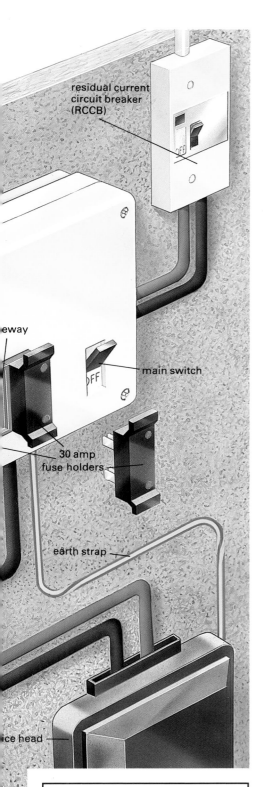

residual current
circuit breaker
(RCCB)

...eway

main switch

30 amp
fuse holders

earth strap

...ce head

MINIATURE CIRCUIT BREAKERS

You can see instantly if an MCB has tripped – the coloured button shoots out of the plug-in holder, or the switch clicks to the 'off' position. The button or switch cannot be set to 'on' until the fault which caused the MCB to trip in the first place has been fixed.

If an MCB trips persistently for no apparent reason, it may be faulty: simply unplug it and swap it with another of the same rating (this is marked on the body, and sometimes indicated by the colour of the button or switch). If the fault transfers to the other circuit, the MCB is definitely faulty. Buy and fit a replacement MCB of the correct rating.

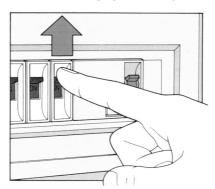

Reset a tripped MCB by pressing in the coloured button or returning the switch to 'on'.

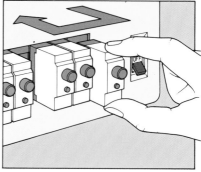

A faulty MCB may keep tripping. Swap it with another one of the same rating to test.

CARTRIDGE FUSES

Many modern consumer units have fuse holders that take a cartridge type fuse. These are generally less fiddly to repair than rewirable fuses, though some types of holder make the job trickier than it needs to be.

The fuse holders are rated differently according to the circuits they protect, and accept only the correct size replacement cartridges (except for 15 and 20 amp fuses, which are interchangeable). This makes it almost impossible to fit a replacement fuse of the wrong rating.

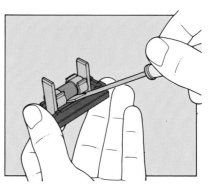

1 Unplug the fuse holder from the consumer unit. With most designs the fuse simply clips into the holder: prise it out with a small screwdriver.

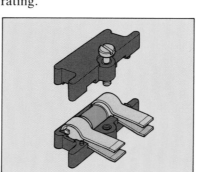

One type of fuse holder is in two parts, held together by a screw. Separate the parts, then remove the cartridge from its hooped terminal clips.

2 Clean up the metal clips with a little wet and dry paper if they look dirty. Clip in the new fuse cartridge, refit the holder and turn on the power.

THE MAIN FUSE

The main fuse protects the mains supply from serious fire or flood damage. It is very rare for it to fail.

You are not allowed to fix the main fuse yourself. The box in which it is fitted (the *service head*) is sealed with a special wire tag to prevent tampering, so call the supply company if you suspect it has blown.

CHECKING PLUG FUSES

Cartridge-type fuses are found not only in plugs, but also adaptors, extension blocks, triple sockets, shaver points and fused connection units (FCUs). Dimmer switches may have miniature cartridge fuses to protect their sensitive circuitry.

FCUs and sockets

FCUs have a fuse in a holder mounted on the faceplate.

Triple sockets (which are prone to overloading) have a similar arrangement to protect the fixed circuit wiring. So do shaver points, though in this case the fuse may be the miniature 1 or 2 amp type.

On some designs the fuse carrier is a tight push-fit in the faceplate and can be prised out with a screwdriver. The other common fitting has a small securing screw.

Adaptors and extensions

The fuses in plug adaptors are usually in a red-coloured holder alongside the plug pins: lever out the holder with a screwdriver. Extension blocks are similarly equipped, though here the holder may be held by a screw.

More rarely, the fuse is found inside the adaptor casing. Undo the screws holding the casing together and very carefully pull it apart: you'll find the fuse clipped into a small circuit board.

Built-in fuses

Dimmer switch fuses (where fitted) are either hidden in the casing or fitted in a pull-out holder.

To gain access to the body-mounted type, turn off the power at the main switch and remove the appropriate lighting circuit fuse. Undo the two fixing screws and pull the entire switch away from the wall. Make a note of where the wires go, release them and remove the switch.

On some designs the fuse is concealed under a plastic insulating cover which must first be unclipped. Double dimmer switches have two fuses, one of which may be difficult to reach since access to it is blocked by the other dimmer control – it is just possible to ease it out with a small screwdriver.

Appliance fuses are normally fitted to a screw-in holder near the flex outlet. If the internal fuse blows, the symptoms are the same as if the plug fuse blows, so check this too if the plug fuse appears to be sound.

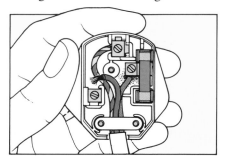

1 Before testing a plug fuse, check that none of the wires have come adrift and watch out for chafed insulation. The flex should be tight in the grip.

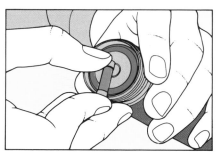

2 Test the fuse by holding one end against the metal part of a torch casing, and the other end to the battery. The torch will light if the fuse is sound.

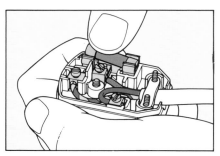

3 Make sure you replace a fuse with one of the correct rating. Clean the terminals, then push the fuse into its clips and refit the top of the plug.

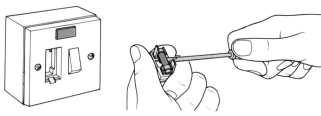

The fuse holder in FCUs and adaptors can usually be prised out with a screwdriver.

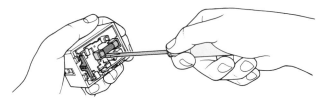

Some adaptors have a fuse mounted on a small circuit board inside the casing.

On double dimmer switches, one of the fuses is obscured – ease it out carefully with a screwdriver.

TYPES OF CARTRIDGE FUSE

CONSUMER UNIT CARTRIDGE FUSES (marked 'BS1361')

Rating	Colour
5 amp	White (lighting circuits)
15 amp	Blue (single appliances up to 3kW)
20 amp	Yellow (single appliances up to 4.5kW)
30 amp	Red (power and cooker circuits)
45 amp	Green (cooker circuits over 12kW)

FUSE WIRE RATINGS
Rewirable fuses have three ratings instead of five:

5 amp	(lighting circuits)
15 amp	(single appliances up to 3kW)
30 amp	(power and cooker circuits)

PLUG CARTRIDGE FUSES (marked 'BS1362')

Rating	Colour
2 amp	Black
3 amp	Red
5 amp	Black
10 amp	Black
13 amp	Brown

NB 3 amp and 13 amp are the most commonly used ratings; only fit fuses of other ratings where specified by the appliance manufacturer.

MINIATURE CARTRIDGE FUSES (marked 'BS646')

Rating	Colour
1 amp	Green
2 amp	Yellow
3 amp	Black
5 amp	Red

REPLACING A SOCKET OUTLET

Socket outlets are generally reliable, and the most common reason for changing one is because the old one is damaged. Stiff or broken pin contacts make plugging in difficult and place a strain on other components, while cracks or holes in the body of the socket are highly dangerous – especially if any of the terminals are exposed. Don't use a socket in such a condition.

Swapping an old, damaged socket for a new one of the same size and type is among the easiest of all electrical jobs. But if the socket is only a single one, you might consider taking the opportunity to replace it with a double – or even triple – outlet socket. In most cases this can be done quite safely (see overleaf), and is a lot better than relying on adaptors or extension leads.

.... Shopping List ....

Replacement socket *faceplates* – the part containing the contacts and terminals – come in standard sizes with a choice of single, double or triple outlets. They may be switched (safer) or unswitched, and the triple type has a built-in 13 amp fuse to guard against overloading.

Faceplates are the same, whether the socket is flush or surface mounted, and so too are the screw positions. See overleaf if you need to replace a surface mounted backing box as well.

Socket styles
Socket designs vary in detail between makes, and you may have to shop around to get an exact match. If you are replacing several sockets to blend in with a new decorative scheme, there is a choice of bright coloured plastic, brass or aluminium (the last two in 'period' styles) as well as traditional white plastic. However, metal faceplates **must** go on a metal backing box (see overleaf).

Tools and materials: Electrical screwdrivers, trimming knife, green and yellow PVC sleeving (maybe), a small length of 2.5mm^2 PVC sheathed cable (maybe).

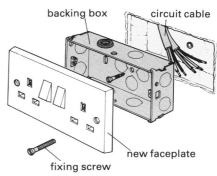

backing box   circuit cable

new faceplate

fixing screw

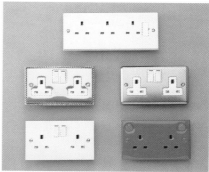

Trade tip

Life savers
❝ If you're replacing a socket that's regularly used to supply garden appliances, think about fitting one with a built-in residual current circuit breaker (RCCB). They're more expensive than plug-in RCCBs, but can't be mislaid or forgotten.

RCCB sockets are designed to fit a standard double socket backing box, so a single socket box must be converted to suit. ❞

REMOVING AND REPLACING

Removing an old socket faceplate is a simple job, since it's only held by two screws and the mains wiring. The procedure is the same for both single and double sockets.

Before you start . . .

■ Switch off the power at the consumer unit and isolate the appropriate circuit fuse or MCB.

■ Plug in an appliance to check there's no power to that socket.

Checking for damage Remove the faceplate and check the cable(s) for damage and correct fitting.

■ On metal backing boxes there should be rubber grommets where the cable(s) enter to stop the PVC sheathing chafing. If there aren't any, it's advisable to remove the backing box and fit some.

■ The earth cores should all be insulated with individual lengths of green and yellow PVC sleeving. Again, if they have been left bare, fit new sleeving before you fit the socket faceplate.

■ Sockets on metal backing boxes must now have a 'flying earth' – a direct connection between the earth terminal on the faceplate and the box itself. You can make the connection using a length of earth wire cut from ordinary 2.5mm² circuit cable, but don't forget to cover it with a length of green and yellow PVC sleeving.

Fitting the new faceplate

The terminals on the new faceplate will be clearly marked: the red wires go to 'L' or 'live', the black wires to 'N' or 'neutral', and the sleeved earth wires to the terminal marked with an earth symbol (see step 4 below).

Where there is more than one cable, check that all the bared ends are pushed fully home in their terminals (see top inset, step 4) before tightening the screws.

1 Undo the faceplate mounting screws, and on a flush socket, cut around any decorations using a trimming knife. Gently ease the socket away from the backing box.

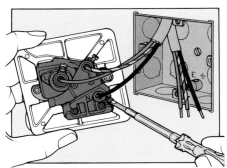

2 Use an electrical screwdriver to slacken off each of the cable terminal screws in turn. Pull the wires free and discard the old faceplate.

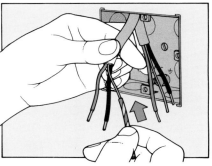

3 Check the condition of the cable(s). All earth wires should be sleeved, and a metal backing box should have rubber grommets at the entry points.

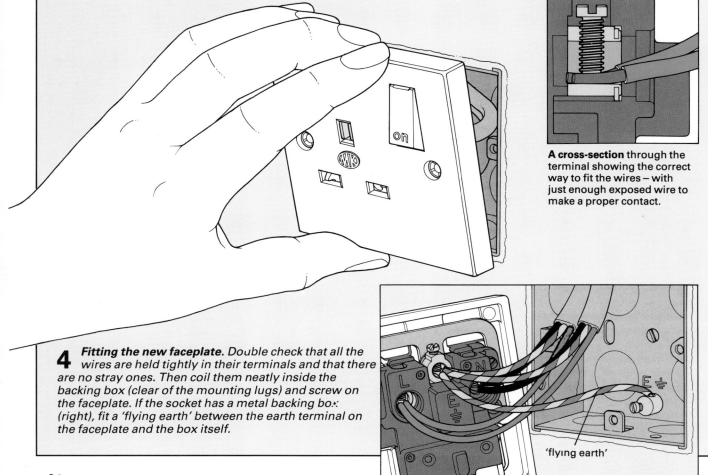

A cross-section through the terminal showing the correct way to fit the wires – with just enough exposed wire to make a proper contact.

4 **Fitting the new faceplate.** Double check that all the wires are held tightly in their terminals and that there are no stray ones. Then coil them neatly inside the backing box (clear of the mounting lugs) and screw on the faceplate. If the socket has a metal backing box (right), fit a 'flying earth' between the earth terminal on the faceplate and the box itself.

'flying earth'

SOCKET CONVERSIONS

There are several ways to convert a single socket to a double or triple, depending on whether you want the new socket to be surface mounted or flush with the wall. But first, check that it's safe to do so.

Surface mounting is by far the easiest option, though the end result won't be as neat as a flush socket, and will be more vulnerable to accidental knocks.

If the existing socket is surface mounted, simply exchange the old single backing box for a larger new one (see Shopping List). But if the old socket is flush mounted, you have the option of fitting a slimmer *conversion box* which screws straight on to the old recessed box (some standard boxes also have mounting holes for fitting this way).

Flush mounting is neater, but involves a lot more work since the hole in the wall must be enlarged to take a new backing box.

Cutting a hole is much easier on a hollow wall than on a masonry one, but in both cases some making good will be needed after the backing box is installed.

> ### ⚠ Party walls
> Don't try to fit extra flush-mounted socket outlets into the party wall of a timber-framed house. This could reduce the effectiveness of the thermal and sound insulation in the cavity.

IS IT SAFE TO CONVERT?

■ Single sockets forming part of a main power circuit, whether ring or radial, can safely be converted to either a double or triple socket.

■ A single socket wired on a **spur** off a main circuit can be converted to a double (but not a triple) socket so long as it is the only socket on that spur.

■ A single socket which shares a spur with another socket (or sockets) – an arrangement now banned by the IEE wiring regulations – cannot be converted. Neither can any of the other sockets on that spur.

Finding out whether a socket is on a spur or part of a main circuit isn't always easy. The number of cables entering the back of the socket gives some useful clues (see chart). But to confirm your suspicions you may have to open up the neighbouring sockets and check these too (spur-wired sockets are nearly always close together). If you are still left in any doubt, don't hesitate to consult an electrician.

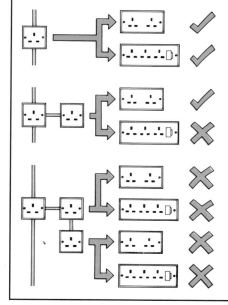

Number of cables in socket	Could be	Safe to convert
Three	Socket on a ring or radial main with a spur connected	Yes
Two	Socket on a ring main	Yes
	Socket on a radial main	Yes
	First of two sockets on a spur	No
One	Last socket on a radial main	Yes
	Only socket on a spur	Yes
	Second socket on a spur	No

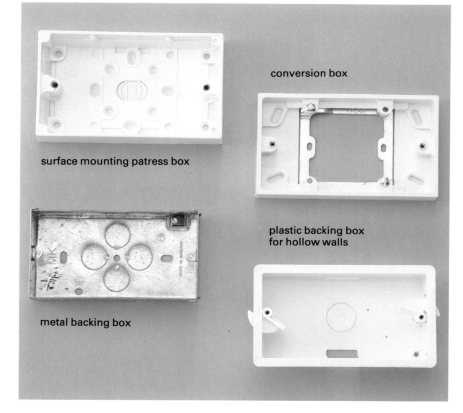

surface mounting patress box

metal backing box

conversion box

plastic backing box for hollow walls

.... Shopping List ....

Having decided how to convert the socket, buy a new backing box to suit. All types of backing box come in standard sizes, matched to the standard sizes for double and triple sockets.

For surface mounting choose between a plastic *patress box* that screws straight to the wall, or a *conversion box* that screws to an existing recessed backing box.

For flush mounting there are two types of recessed backing box. *Metal backing boxes* are for masonry walls, or hollow walls where the socket coincides with a vertical stud that the box can be screwed to. *Hollow wall boxes* are designed to clip into a hole cut in the plasterboard – they don't need to be screwed in place.

Tools and other materials: Screwdrivers, trimming knife. Maybe also: electric drill and bits, wallplugs, padsaw, hammer and bolster, goggles, rubber grommets.

SURFACE MOUNTING

If you're replacing an existing surface mounted patress box, simply undo the fixing screws and ease it away from the wall. The new box will have a variety of blanked-off mounting holes, so you should be able to match the existing screw holes. If not, drill and plug new ones to accept the new box.

Knock out the mounting holes and cable entry hole(s), feed the cable(s) into the box, and screw in position. Fit the new faceplate as described previously.

If you're fitting a conversion box over an old flush-mounted backing box, this screws straight on. If screws aren't supplied, you can use the mounting screws from the old socket.

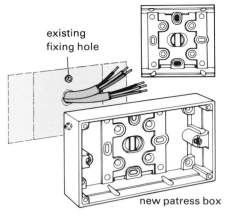

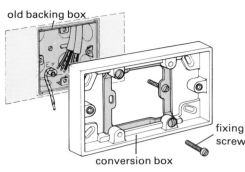

Double patress boxes *have a variety of blanked-off mounting holes, so you should be able to match the existing screw holes. If not, drill and plug new ones.*

Conversion boxes *simply screw over the old flush-mounted backing box, but check first that this is securely fixed to the wall and has a 'flying' earth.*

FLUSH MOUNTING

Fitting a flush mounted backing box is fairly easy on hollow walls.
■ Undo the two screws holding the old backing box and ease it out.
■ Mark the position of the new box on the wall with a pencil. Keep the box central on the existing hole if possible.
■ Cut out the marked area with a trimming knife and steel rule or a padsaw. Sand down the edges.
■ Remove the appropriate

knockout from the new box, fit a rubber grommet, feed in the cable(s), and push the box into the hole. Plastic boxes clip in, metal ones screw to the stud.

On a masonry wall, unscrew and prise out the old backing box, then offer up the new box and mark the outline on the wall.

The easiest way to enlarge the recess is to drill a series of holes each side of it with a 10mm masonry

bit, then square up the edges with a club hammer and bolster. Wear goggles or safety spectacles for this, and take great care not to drill or chisel near the cables; if necessary, loosen the plaster around them and pull them clear as you go.

Remove a knockout for the cable(s), fit rubber grommet(s) and feed in the cable(s). Try the box for fit; level it up if need be using slips of wood, then screw it in place.

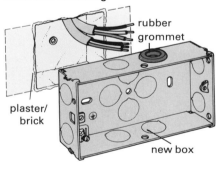

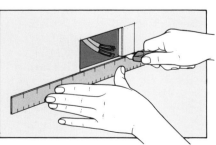

On a hollow wall, *cut back the plasterboard to the size of the new box, then clip the box in place or screw it to the stud depending on the type.*

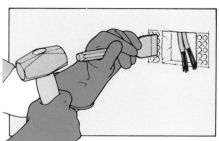

On a solid wall *enlarge the recess by drilling a series of holes and then trimming with a bolster. Take great care not to damage the cables.*

PROBLEM SOLVER

Drilling holes
Since the electricity will be off, you could run into problems if you find you have to drill the wall during a socket conversion. The simple answer is to use a cordless drill. But if you don't own one, the only alternative is to run the drill via an extension lead from a socket on a cooker control unit or another power circuit.

Before you turn the electricity back on to do this, **make certain that the circuit supplying the socket you are working on is isolated** – in other words, that you've removed the appropriate fuse (or tripped the appropriate MCB). Double check by testing the socket with an appliance – even if this means you have to put everything back as it was.

ADDING EXTRA SOCKETS

It's a sad fact that most homes still don't have enough socket outlets to cope with the demands of modern living. One answer is to use adaptors or extension leads, but these can be dangerous as well as inconvenient, and they are certainly no good for powering fixed appliances such as an extractor fan or waste disposal unit. The only satisfactory solution is to fit extra sockets or fused connection units (FCUs) exactly where you need them, so that leads are kept short and outlets can't get overloaded.

What the job involves
There is no difference between extra sockets and FCUs as far as wiring into an existing power circuit is concerned, so you can regard the following instructions as interchangeable. Basically, there are two ways to do the job:
■ Wire the new socket from a junction box fitted in the power circuit cable at a convenient point.

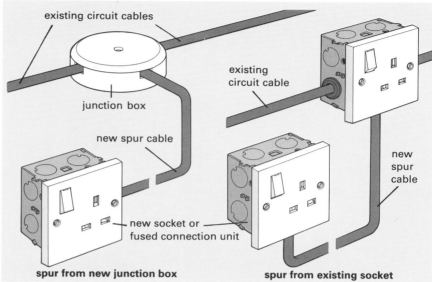

spur from new junction box **spur from existing socket**

■ Wire the new socket from a socket already on the power circuit.
Sockets wired in this way are known in the trade as *spurs*.
Physically running the cables, fitting the new parts and connecting everything up are straightforward

New sockets *can be wired from a junction box (left) or socket.*

enough jobs. But first, you must know whether or not it is safe to add a socket within the constraints laid down by the Wiring Regulations.

IS IT SAFE TO ADD A SOCKET?

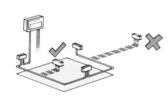

20 amp radial circuit

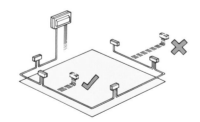

30 amp radial circuit

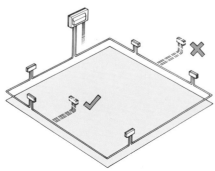

30 amp ring circuit

The Wiring Regulations impose restrictions on adding sockets, to stop circuits becoming overloaded.

How many sockets?
The total number of sockets you can have on a power circuit is limited only by the floor area which that circuit serves. However, the total number of sockets wired as spurs must not exceed the number of sockets on the main circuit.

As the table below shows, the area limit depends on the fuse rating of the circuit, and on whether it is wired on the modern *ring* or older *radial* system. The table also shows what size PVC sheathed cable to use for wiring up the new socket.

Type of circuit	Max. floor area	Cable size
30 amp ring (red fuse)	100sq m (120sq yd)	2.5mm²
30 amp radial (red fuse)	50sq m (60sq yd)	4mm²
20 amp radial (yellow fuse)	20sq m (24sq yd)	2.5mm²

Wiring from a junction box
So long as you don't exceed the specified floor area, you can fit a junction box anywhere on the circuit cable. However, you may only fit

one new socket per box, and you can't fit a junction box in a cable which has already been spurred off the main circuit.

Wiring from a socket
Here, everything depends on where the socket from which you want to take the spur stands in the circuit.
■ You **can** wire from any socket forming part of a main ring or radial power circuit so long as no other socket has already been wired from it in this way.
■ You **cannot** wire from a socket which is itself a spur, or which is a spur off a spur (no longer allowed by the Wiring Regulations).
Telling the difference between a circuit socket and a spur socket is not always easy, but you must know before going ahead. The checking process is described in Problem Solver overleaf, but if you are still in any doubt, the job should be left to a qualified electrician.

What sort of socket?
The socket you add as a spur can be a single socket, a double socket, or a single FCU. You cannot add a triple socket as a spur.

When is a spur not a spur?

Establishing whether a socket forms part of a main power circuit or is a spur off that circuit is not always as straightforward as it seems. There is no 'quick' way to find out; the only reliable method is to go through the checking sequence given here, eliminating the possibilities along the way until you arrive at the logical conclusion.

CHECK 1 How many circuit cables enter the socket? Turn off the mains and unscrew the socket faceplate to check. The table on the right lists the possibilities, of which the first two can be eliminated straight away.

No. of cables	Socket could be. . .	Suitable for connection
3	Part of a ring circuit with a spur wired off it	No
3	Part of a radial circuit with a spur wired off it	No
2	Part of a ring circuit	Yes
2	Part of a radial circuit	Yes
2	A spur feeding another spur (no longer allowed)	No
1	A spur off a ring or radial circuit	No
1	A spur off another spur	No
1	The end of a radial circuit	Yes

CHECK 2 Which power circuit is the socket on? Plug in a table lamp, then remove the power circuit fuses (or trip the relevant MCBs) at the consumer unit one by one until the socket goes dead. Repeat the test for the two neighbouring sockets, as you may need to know if they share the same circuit. Always turn off at the mains when removing and replacing fuses.

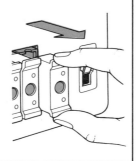

CHECK 3 Check the colour coding of the fuseholder/MCB; see how many circuit cables enter the fuseway. (If you remove the consumer unit cover turn off at the mains first.)

Colour coding	Cables in fuseway	Type of circuit
red	2	ring 30 amp
red	1	radial 30 amp
yellow	1	radial 20 amp

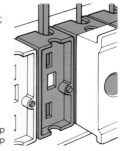

CHECK 4 RING CIRCUITS
■ If the socket is on a ring circuit and has only one cable entering it, then it is a spur and cannot be connected to.
■ If the socket has two cables, it could be part of the main ring, or else a spur feeding a spur (now banned).

To find out which, turn off the mains and connect a continuity tester between the live cores of the two cables. Continuity means the socket is on the ring and is safe to connect to; no continuity means it is a spur feeding a spur and can't be connected to.

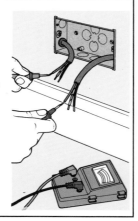

CHECK 5 – RADIAL CIRCUITS
Depending on whether one or two cables enter the socket, it could still be any of the possibilities shown in the diagram. However, by checking how many cables enter neighbouring sockets which you know are on the same circuit, you should be able to narrow down the possibilities to one.

For example, if the socket has two cables, while one of its neighbours has three and the other one, you know that the first socket is a spur feeding a spur and can't be connected to.

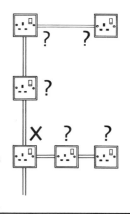

....Shopping List....

The new socket or FCU can be surface mounted – which involves less work – or flush mounted within the wall, which looks neater and will protect it from knocks.
For surface mounting, mount the socket on a matching *pattress box*. Leave the cable exposed, using matching size *cable clips* to secure it, or run it in *plastic mini-trunking* glued to the wall with *impact adhesive*. In a workshop or garage, use heavy-duty *metal clad* sockets and run the cable in impact resistant *plastic conduit*.
For flush mounting in a solid wall, mount the socket in a *metal backing box* and recess the cable in a channel (chase) cut through the plaster. Encase the cable in plastic conduit where extra protection is required.

Flush mounting in a hollow wall can be done in the same way, cutting a hole in the plasterboard and screwing the backing box to a frame member. Alternatively, you can use a special *hollow wall box* which clips to the plasterboard.
For circuit cable connection other than at a socket, buy a round or rectangular *30 amp 3 terminal junction box* (see page 42).

Other materials
Whatever the mounting method, you will need:
PVC sheathed twin and earth cable of the appropriate size for the circuit being wired into. Estimate on the generous side – you are bound to need more than you imagine.
Green and yellow PVC sleeving for insulating the bare earth wire at connection points.
Rubber grommets for protecting cable entering metal boxes.
Making good materials – repair plaster and DIY skimming plaster.
Tools checklist: Screwdrivers, electric drill and bits, wire strippers, trimming knife, hammer, club hammer and bolster (maybe).

Other tools may be required if you need to gain access to an existing circuit cable.

FLUSH MOUNTING A SOCKET

Mount the new socket or FCU and run the cable before connecting into the existing circuit, so that the time the power is turned off is kept to a minimum. The safe height for a socket is a minimum of 150mm (6") above the floor or work surface, but keep all sockets well away from a kitchen sink.

For an FCU wired to a fixed appliance with cable, it's advisable to cut the chases for both circuit and appliance cables while mounting the backing box.

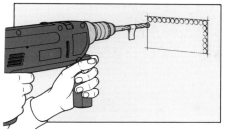

1 *On a solid wall,* mark around the backing box in pencil, then chain-drill a series of holes to the same depth as the box using a masonry bit.

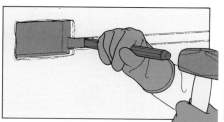

2 Chop out the hole for the backing box, plus a chase for the cable, using a hammer and bolster. Try the box in position, level with slips of wood, then drill and plug fixing holes.

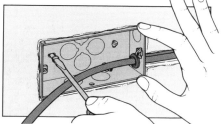

3 Remove the appropriate cable knockout on the box with a screwdriver, fit a rubber grommet, then feed in the new cable. Refit the box in the hole and screw in place.

SURFACE MOUNTING

1 Position the pattress and mark the screw holes. (On a hollow wall try to align the holes with a stud.) Drill the holes then fit wallplugs if necessary.

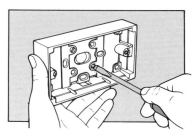

2 Carefully break away the plastic over the appropriate cable entry and remove any sharp edges. Secure the pattress to the wall with No.8 woodscrews.

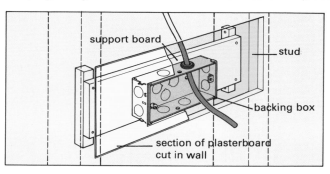

On a hollow wall, the traditional flush mounting method is to mark the box position, cut out the hole using a trimming knife, then cut out the entire panel between two studs. Screw the box to a piece of timber or plywood and nail this between the studs so that the box sits flush with the surface. Replace the plasterboard panel after feeding in the cable as described above.

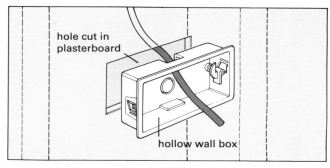

A hollow wall box makes the job a lot easier, since there is no need to cut out a large access panel or fit a supporting frame member.

Mark and cut a hole for the box, then feed in the new circuit cable. Afterwards, slot the box into the hole and push until the clips engage, or push and rotate the clips, depending on the design.

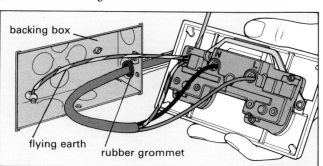

Connect the prepared cable to the new socket faceplate – red to L (live), black to N (neutral) and the earth wire to Earth. For a metal box, connect a separate length of sleeved earth wire between the earth terminal on the faceplate and the terminal on the box. Double check all the wires are tight, then carefully coil the wires into the backing box and screw on the faceplate. Keep wires away from the fixing lugs on metal boxes.

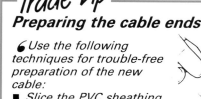

Trade tip
Preparing the cable ends

Use the following techniques for trouble-free preparation of the new cable:
- *Slice the PVC sheathing lengthways along the bare earth wire, fold back by the required amount and trim off.*
- *Strip about 12mm (½") of insulation from the red and black wires. Then cut a length of green and yellow PVC sleeving and slip over the earth wire.*

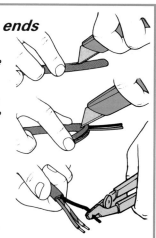

CONNECTING TO THE CIRCUIT – AT A SOCKET

Run the cable as far as the socket, then turn off the mains and remove the appropriate circuit fuse. Unscrew the socket faceplate, and disconnect the existing circuit cable wires from their terminals.

For a surface mounted box, simply knock out another cable entry in the pattress box, then feed in the new cable and trim to leave about 150mm (6") remaining.

For a flush mounted box, unscrew the box from the wall. Work out how the new cable is to enter, and if necessary chop out the remaining section of chase in the plaster, taking care not to damage the existing cables. Then knock out a fresh entry hole in the box, fit a rubber grommet, and feed in the cable. Fit the box in the wall, and trim back the new cable as above.

Connecting the wires

Strip back the outer PVC sheathing on the new cable by the same amount as the other cables, then remove 12mm (½") of insulation from the red and black cores.

■ Group the red cores together and connect to the L (live) terminal.

■ Group the black cores together

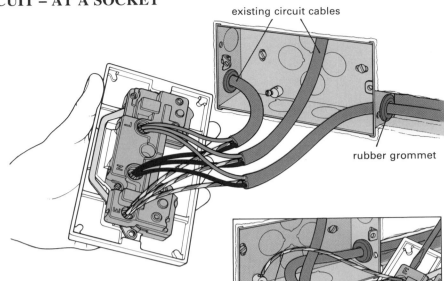

existing circuit cables

rubber grommet

Run in the new cable and reconnect the grouped cable cores to the faceplate as shown. On a flush metal box, make sure there is a 'flying earth' link (right).

'flying earth'

and connect to the N (neutral) terminal.

■ Slip green and yellow PVC sleeving over the bare earth wire on the new cable, then group the earth wires together and connect to the Earth terminal on the faceplate.

On a flush box, check that there is a 'flying earth' connection between the Earth terminal and the box.

Double check that all the wires are held securely in their terminals, then gently coil them back into the box and replace the faceplate.

CONNECTING TO THE CIRCUIT – AT A JUNCTION BOX

Connecting at a junction box means exposing and identifying the circuit cable, then finding a suitable point to break into it. The direction taken by the cable as it enters and leaves nearby sockets should give a clue to its position; be sure to turn off the mains before you check.

Exposing the cable itself may involve considerably more exploratory work, so be prepared to lift floorboards – or to cut access holes, if it runs through a hollow wall. The junction box can be screwed to any

nearby joist or stud. Note however, that junction boxes cannot be buried in plaster.

With the cable exposed, double-check that it is the correct power circuit cable by tugging on it and registering the movement at a nearby socket. (The socket must not, of course, be a spur.) You can then run in the new cable and connect to the junction box as shown.

With a round box, sever the circuit cable and prepare the ends in the normal way. Connect the three

groups of wires to their appropriate terminals, not forgetting to sleeve the bare earths.

With a rectangular box, strip away 40mm (1½") of the circuit cable outer sheath and bare the red and black cores to leave them exposed but undamaged. Cut through the earth core, sleeve the two parts, then lay the cable in the box as shown. Complete the job by preparing the end of the new cable and connecting the groups of wires to their appropriate terminals.

RECTANGULAR JUNCTION BOX

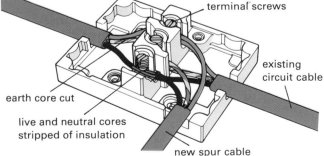

terminal screws

existing circuit cable

earth core cut

live and neutral cores stripped of insulation

new spur cable

On a round junction box, sever the circuit cable and prepare the ends, then group the three sets of cable wires and connect to the appropriate terminals. Afterwards, refit the screw cover so that unused cable entry holes are blanked off.

ROUND JUNCTION BOX

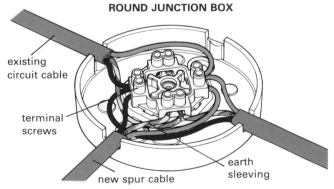

existing circuit cable

terminal screws

new spur cable

earth sleeving

On a rectangular junction box, only the circuit cable earth core needs to be cut (so that it can be sleeved). The live and neutral cores can simply be bared, then laid in their terminals ready for connection to the new spur cable.

RUNNING CABLES

Whether you're adding a new socket or rewiring an entire circuit there are two main ways to run the cable: **Surface mounting** is the easiest, but not necessarily the neatest, choice. **Concealing** the cable in a wall, floor or ceiling results in a more professional job but is more work.

There are a number of different options for both surface and concealed cable runs. Consult the chart on the right and page 46 for the one best suited to the job.

Where to run cable

You can't just run cable anywhere – there are a number of wiring conventions you should follow to protect both yourself and anyone who has to work on the wiring later.

■ Run cable in vertical or horizontal lines from fitting to fitting. This makes tracing the circuit easier. Never run a cable diagonally – it may save on cable, but when concealed gives no clue to its position, so someone may accidentally drill or nail into it.

■ Run hidden cables within 150mm (6″) of a floor, ceiling or adjoining wall; then to the fitting.

■ Plan ahead; if you're having a new concrete floor laid or a wall built, have any cables and conduit put in at the same time. This avoids a lot of mess and expense later.

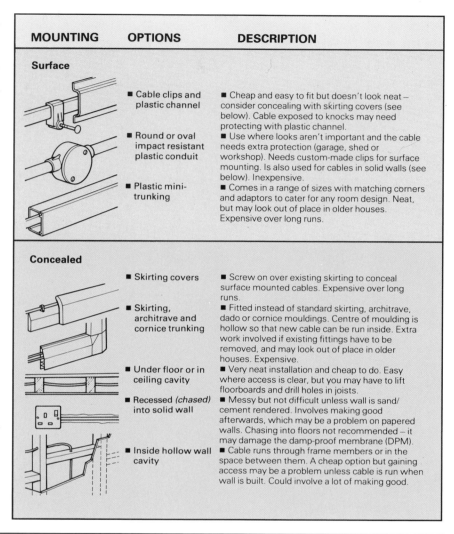

MOUNTING	OPTIONS	DESCRIPTION
Surface	■ Cable clips and plastic channel	■ Cheap and easy to fit but doesn't look neat – consider concealing with skirting covers (see below). Cable exposed to knocks may need protecting with plastic channel.
	■ Round or oval impact resistant plastic conduit	■ Use where looks aren't important and the cable needs extra protection (garage, shed or workshop). Needs custom-made clips for surface mounting. Is also used for cables in solid walls (see below). Inexpensive.
	■ Plastic mini-trunking	■ Comes in a range of sizes with matching corners and adaptors to cater for any room design. Neat, but may look out of place in older houses. Expensive over long runs.
Concealed	■ Skirting covers	■ Screw on over existing skirting to conceal surface mounted cables. Expensive over long runs.
	■ Skirting, architrave and cornice trunking	■ Fitted instead of standard skirting, architrave, dado or cornice mouldings. Centre of moulding is hollow so that new cable can be run inside. Extra work involved if existing fittings have to be removed, and may look out of place in older houses. Expensive.
	■ Under floor or in ceiling cavity	■ Very neat installation and cheap to do. Easy where access is clear, but you may have to lift floorboards and drill holes in joists.
	■ Recessed (chased) into solid wall	■ Messy but not difficult unless wall is sand/cement rendered. Involves making good afterwards, which may be a problem on papered walls. Chasing into floors not recommended – it may damage the damp-proof membrane (DPM)
	■ Inside hollow wall cavity	■ Cable runs through frame members or in the space between them. A cheap option but gaining access may be a problem unless cable is run when wall is built. Could involve a lot of making good.

SURFACE MOUNTING WITH CABLE CLIPS

Run the cables along room features such as skirtings and architraves – they will be less obvious and better protected from accidental damage. On skirting, clip to the top edge. Make sure the clips are the correct size for the cable and correctly spaced to prevent sagging.

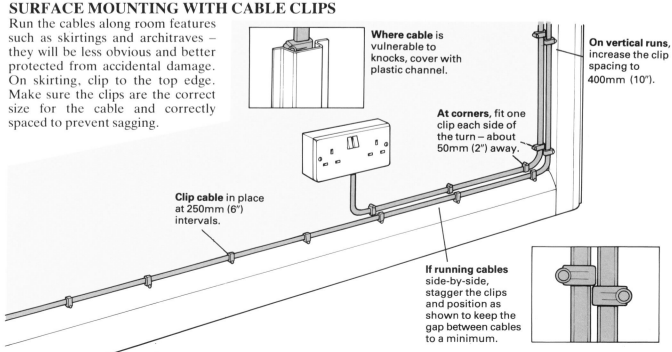

Where cable is vulnerable to knocks, cover with plastic channel.

On vertical runs, increase the clip spacing to 400mm (10″).

At corners, fit one clip each side of the turn – about 50mm (2″) away.

Clip cable in place at 250mm (6″) intervals.

If running cables side-by-side, stagger the clips and position as shown to keep the gap between cables to a minimum.

FITTING MINI-TRUNKING

Plan out the run before fixing: you'll find it easiest to mark the positions of junctions, corners and power socket or light switch boxes using the fittings themselves as templates. You can then measure off straight lengths direct.

Cut the channel and its cover with a junior hacksaw, smoothing off any burrs with sandpaper or a rasp. The easiest way to fix it is with impact adhesive, but on some surfaces (eg wallpaper), it's better to nail. In this case, fix at 500mm (20″) intervals, with the last fixings within 100mm (4″) of the end. Some manufacturers also recommend the nail holes are enlarged to an oval to allow for expansion.

Trade tip

Hidden trunking

6 *If you're worried that mini-trunking will stick out from the room decor, use the smallest size that will safely take the cable and run it right up against the skirting and architraves. With luck it will 'disappear' into the moulding.* 9

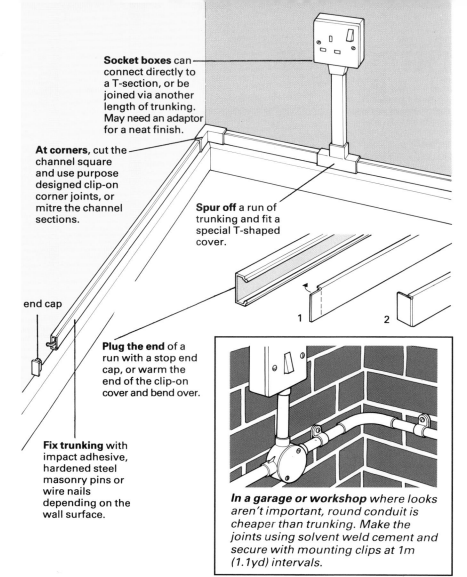

Socket boxes can connect directly to a T-section, or be joined via another length of trunking. May need an adaptor for a neat finish.

At corners, cut the channel square and use purpose designed clip-on corner joints, or mitre the channel sections.

Spur off a run of trunking and fit a special T-shaped cover.

end cap

Plug the end of a run with a stop end cap, or warm the end of the clip-on cover and bend over.

Fix trunking with impact adhesive, hardened steel masonry pins or wire nails depending on the wall surface.

1   2

In a garage or workshop where looks aren't important, round conduit is cheaper than trunking. Make the joints using solvent weld cement and secure with mounting clips at 1m (1.1yd) intervals.

FLOOR OR LOFT WIRING

Running cable under wooden floors or in ceiling cavities is a neat method, but only feasible if you can lift the floorboards or work from an unboarded loft. Unless you are dealing with a ground floor that has free space below, it will be much easier if the cable can run at right angles to the boards – that is to say, parallel with the joists.

In this case, lift boards at each end of the proposed run – and if necessary, at intervals in between. Then use a long thin batten, an electrician's draw tape or a couple of plastic drain rods to feed the cable through.

Routing cable across the joists isn't difficult if there's room underneath – simply clip to each one. But in most cases, the joists have to be drilled or notched, taking care not to weaken the timbers by removing more wood than is necessary.

In lofts, cables must be laid over the top of any roof insulation and clipped to the top surface of each of the joists.

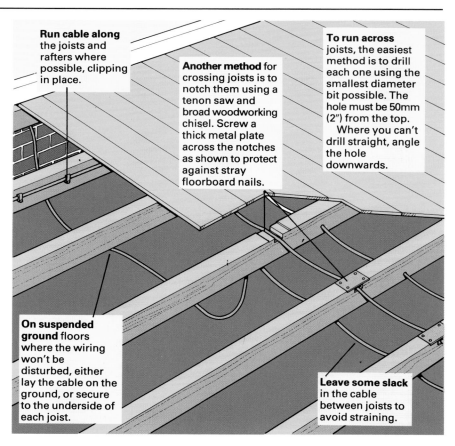

Run cable along the joists and rafters where possible, clipping in place.

Another method for crossing joists is to notch them using a tenon saw and broad woodworking chisel. Screw a thick metal plate across the notches as shown to protect against stray floorboard nails.

To run across joists, the easiest method is to drill each one using the smallest diameter bit possible. The hole must be 50mm (2″) from the top.

Where you can't drill straight, angle the hole downwards.

On suspended ground floors where the wiring won't be disturbed, either lay the cable on the ground, or secure to the underside of each joist.

Leave some slack in the cable between joists to avoid straining.

RUNNING CABLE IN A HOLLOW WALL

To run in from above, lift the upstairs floorboards and find the head plate. Drill a hole above where the fitting is to go, and cut an access hole at the fitting position.

Feed a weighted string down into the wall with the cable attached to the other end, and aim to hook it out through the access hole. If there is a noggin in the way, you must cut a second access hole and drill through this too.

To run in from below, cut an access hole just above skirting level and drill through the sole plate into the floor cavity. Feed the cable in from under the floor, and draw it up on a weighted string fed down from an access hole at the mounting position.

Horizontal runs in hollow walls are best made during building, before the framework is clad with plasterboard. In this case treat the studs as joists; either drill through them, or notch them and cover with metal plates. In an existing wall, cut an access hole at each stud.

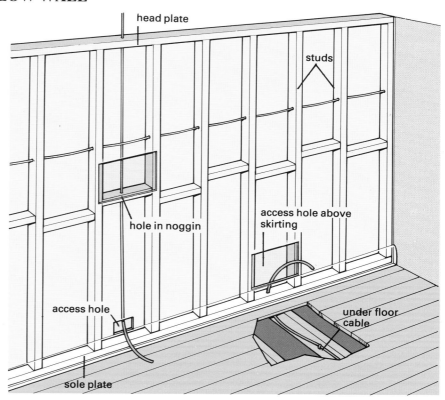

head plate

studs

hole in noggin

access hole above skirting

access hole

under floor cable

sole plate

CHASING IN CABLE

Recessing a cable into a solid wall or floor involves cutting a channel known as a *chase*. PVC sheathed cable can be chased in a wall and plastered over without any extra protection; but if the cable is in a floor, or may need to come out later, it should be run through plastic conduit (see overleaf).

You can hire chase-cutting tools (or drill attachments) but for one-off jobs it's hardly worth it: use a sharp electrician's bolster, cold chisel and club hammer.

Make sure you cut the chase wide and deep enough: 25mm (1″) is sufficient for 2.5mm² cable or smaller, whether exposed or run in conduit; larger cable chases must be substantially deeper. In older walls the plaster coat alone may be thick enough to accommodate the chase; on newer walls, you may have to cut into the masonry too.

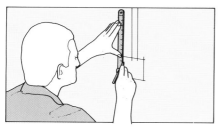

1 Mark the cable run and the area to be chopped away. Make the chase about 25mm (1″) wide for a single run of cable or conduit; 50mm (2″) for a double.

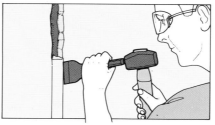

2 Cut away the plaster from the marked area using a hammer and electrician's bolster. Finish off with a smaller cold chisel to get a neat edge.

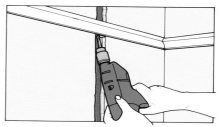

3 Where the run passes behind a skirting or a dado, use a drill with long masonry bit to clear the channel, then finish off with a slim cold chisel.

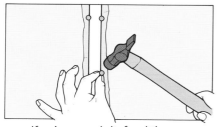

4 If using conduit, feed the new cable through before fitting – it may be difficult to do later. Use galvanized clout nails to hold the conduit in place.

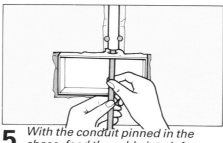

5 With the conduit pinned in the chase, feed the cable into it from one end of the run. Leave plenty for later connection at accessory positions.

6 When you have run in all the new cables, conceal the conduit with filler on short runs and with patching plaster on longer stretches.

CABLE RUNNING ACCESSORIES

IMPACT-RESISTANT PLASTIC CONDUIT

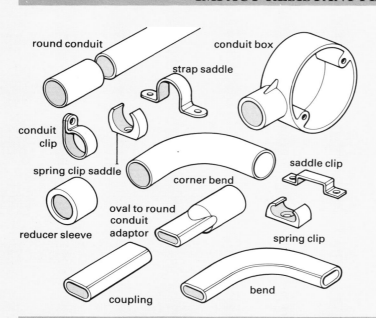

round conduit

conduit box

strap saddle

conduit clip

spring clip saddle

corner bend

saddle clip

reducer sleeve

oval to round conduit adaptor

spring clip

coupling

bend

Conduit is used where cable needs extra protection. Round and oval versions are available, the first in plastic or steel, the second in plastic only.
Round section plastic conduit is usually surface mounted, but may be chased into a wall. The most common size is 20mm which usually comes in 2–3m (7–10′) lengths, but you may also come across 16mm and 25mm. The 20mm size plugs straight into the cable entry of any standard circular conduit (BESA) box or (via an adaptor) into most backing boxes.

The sections are connected with *coupling sleeves* and *bend sections* (which may have access covers), secured with a solvent weld cement. Match 20mm conduit to other sizes with a *reducer sleeve*. Use matching *conduit clips, strap,* or *spring clip saddles* to secure. *Steel conduit* is not suitable for DIY use.
Oval conduit is used for running cables chased into walls or floors. There's a range of sizes, but the ones you're likely to need are 13mm, 16mm and 20mm which may come in 2–3m (7–10′) lengths. Couplers and corners are available for some sizes, and an adaptor permits linking into round section conduit.

MINI-TRUNKING SYSTEMS

Mini-trunking consists of a plastic U-shaped channel with a clip-on cover which glues, nails or screws to the wall. It comes in 3m (10′) lengths that can be joined with *couplings* for longer runs.

A variety of sizes is available, depending on the number and size of cables you want to run. The smallest standard mini-trunking is 16×16mm but there is a range of smaller sizes (sometimes called *communication trunking*) which is very compact and easier to disguise. Although designed for running phone cables, bell wire and hi-fi wiring it can be used to run a single power or lighting cable (up to $2.5mm^2$ or $4mm^2$ depending on the size). For larger cable sizes or double cable runs use mini-trunking. Don't run mains and phone or hi-fi cables in the same trunking; it may cause interference.

Some manufacturers make *twin-compartment* trunking to keep things neat. Accessories to make running the trunking easier include flat, internal and external corners (often called *angles*), T-pieces (for adding a spur) and mounting boxes.

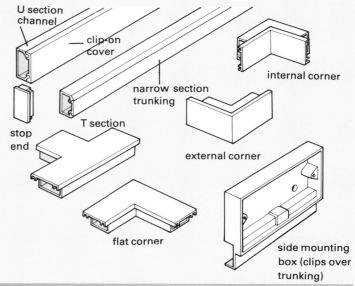

U section channel

clip-on cover

narrow section trunking

internal corner

T section

stop end

external corner

flat corner

side mounting box (clips over trunking)

TRUNKING MOULDINGS

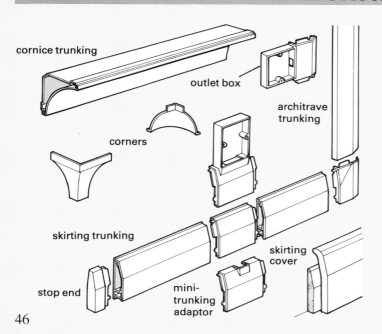

cornice trunking

outlet box

architrave trunking

corners

skirting trunking

skirting cover

stop end

mini-trunking adaptor

Larger versions of mini-trunking are shaped to resemble – and designed to replace – various feature mouldings around the house. In many cases each type can be connected to the others (and to mini-trunking) to build-up an integrated system.
Skirting trunking comes in straight lengths and replaces an existing skirting run. Fitting accessories include *corners, joint plates, stop ends*, adaptors (for connecting into matching architrave or mini-trunking systems) and *outlet boxes* for fitting power sockets. *Dado trunking* is similar but has different shaped clip-on covers.
Architrave trunking is used for continuing the run neatly around door openings and has its own accessories.
Cornice trunking for running cable at ceiling level comes in 3m (10′) lengths with matching end caps, corners, joint covers and mini-trunking adaptors.

Skirting trunking is sometimes confused with **skirting cover kits** which are used to conceal wiring runs clipped to the top of the existing skirting. The cover is deeper than the existing skirting and simply screws or glues on over the top.

ADDING A SHAVER SOCKET

Although you can power an electric razor from a normal socket, using an adaptor plug, this is not always convenient. In a bedroom the sockets are often too low for the flex to stretch easily, and in the bathroom ordinary sockets are not allowed.

The solution is to fit a special shaver point. There are two types:
In bathrooms or anywhere else that gets wet, you must use a *shaver supply unit* to BS3535. These have an *isolating transformer* to separate the socket from the mains supply and prevent any danger of electrocution.
In bedrooms you can normally use a cheaper *shaver socket* (BS4573). These don't have a transformer but are fitted with a 1 amp protective fuse. A bedroom with an en-suite shower or bath counts as a bathroom and you must fit a shaver supply unit unless it is more than 2.5m (8′) away from the water.

Wiring options
■ Unlike other power sockets, you can connect a shaver to the lighting circuit. This is an easy option in a bathroom, where there is unlikely to be a nearby power circuit. There are even combined light/shaver point fittings (see Problem Solver).
■ Spurring the shaver socket off a power circuit is likely to be easier in a bedroom, but the Wiring Regulations require the spur to be wired via a fused connection unit.

For either option, run the cables in any of the normal ways. Shaver sockets fit a standard 25mm deep single socket box, but shaver supply units need a 45mm deep backing box. Fitting this in a solid wall means chopping out, and in thin partition walls there is a danger of going right through. If you think this may happen, use a surface mounting box.

CONNECTING TO A POWER CIRCUIT

Choose the nearest convenient socket outlet to run a spur cable to the shaver point. The details are much the same as running an extra 13 amp socket on a spur (see pages 39-42) – except that the spur cable must run via a fused connection unit (FCU) fitted with a 3 amp fuse. This must be outside a bathroom.

Where there is no suitable socket outlet, the cable can be spurred off a power circuit at any point using a junction box. Again, it must go through a 3 amp fused FCU.

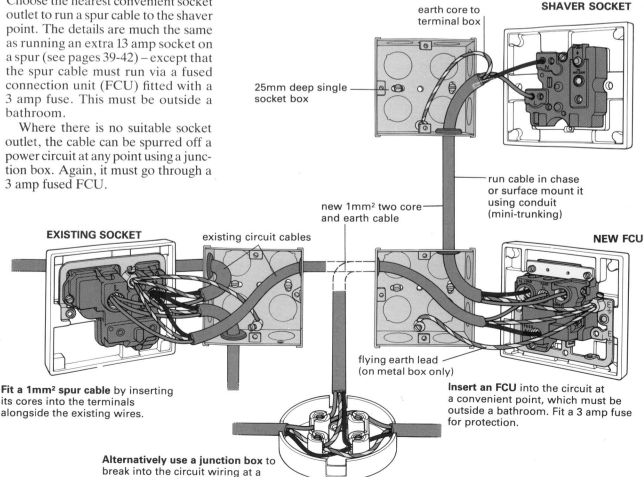

earth core to terminal box

SHAVER SOCKET

25mm deep single socket box

run cable in chase or surface mount it using conduit (mini-trunking)

EXISTING SOCKET

existing circuit cables

new 1mm² two core and earth cable

NEW FCU

flying earth lead (on metal box only)

Fit a 1mm² spur cable by inserting its cores into the terminals alongside the existing wires.

Insert an FCU into the circuit at a convenient point, which must be outside a bathroom. Fit a 3 amp fuse for protection.

Alternatively use a junction box to break into the circuit wiring at a convenient spot.

CONNECTING TO A LIGHTING CIRCUIT

Run the cable from the shaver point up through the ceiling wherever convenient and connect it to the lighting cables above. In all cases it's best to connect to the unswitched supply so that the shaver socket is live all the time. If you wire into the switched side of the circuit the socket will only work when the light is on.

On a loop-in system connect to the terminals in the rose or in the light fitting.

On a junction box system make the connections at the junction box.

If either option proves difficult you can break into the supply cable direct (see Tip).

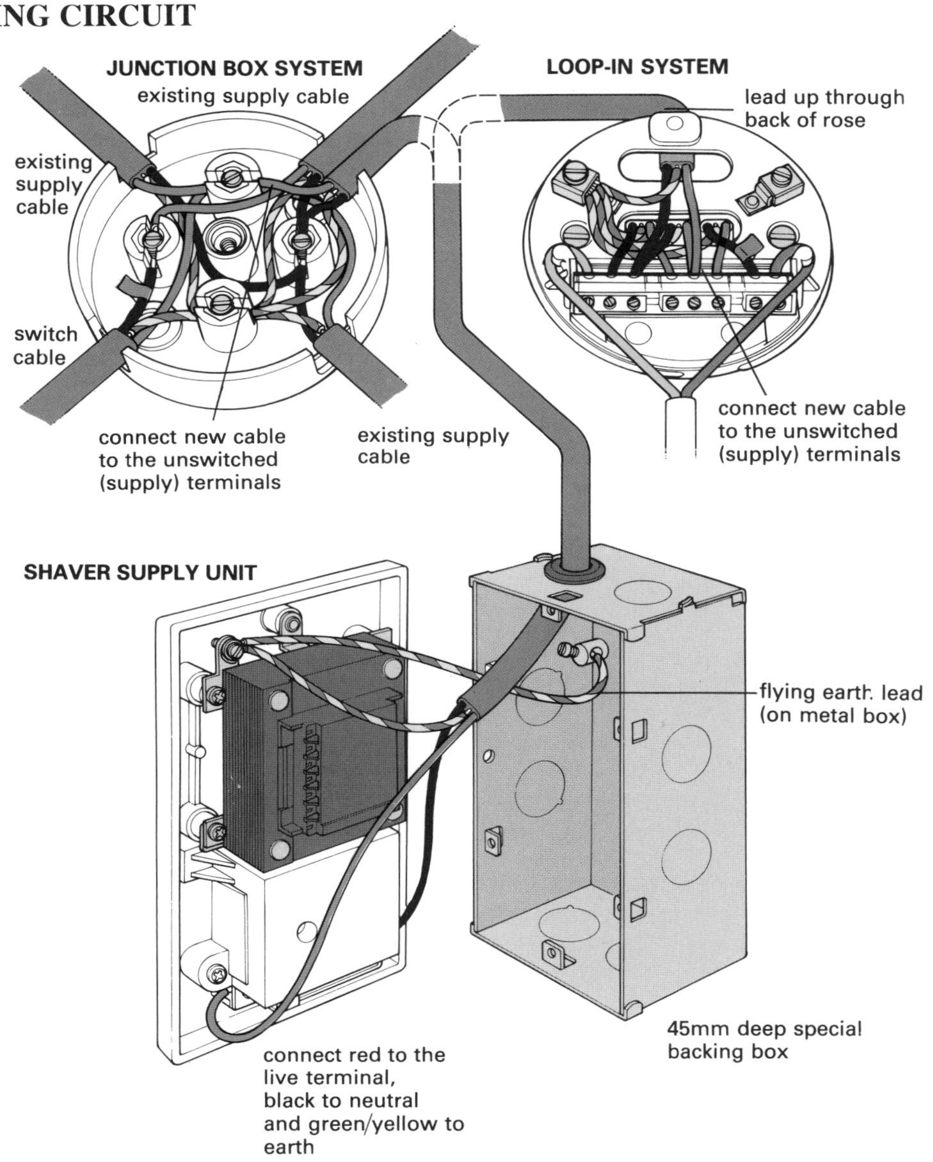

JUNCTION BOX SYSTEM
existing supply cable

existing supply cable

switch cable

connect new cable to the unswitched (supply) terminals

existing supply cable

LOOP-IN SYSTEM

lead up through back of rose

connect new cable to the unswitched (supply) terminals

SHAVER SUPPLY UNIT

flying earth. lead (on metal box)

45mm deep special backing box

connect red to the live terminal, black to neutral and green/yellow to earth

---

Trade tip

The shortest route

❛Connect into the lighting circuit using a junction box if there is no nearby rose or junction box to take the power from.

Cut any nearby supply cable and insert a three-way junction box to reconnect the two ends. Then connect your new cable to the same terminals. ❜

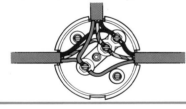

▮ PROBLEM SOLVER

Preserving the decorations

Fitting a shaver point involves a good deal of disruption – especially in a tiled bathroom, where it may be very difficult to get new tiles to match.

However, where there is an existing wall-mounted light an easy solution is to replace it with a combined striplight and shaver socket. The wiring will all be there, so all you need do is drill mounting holes and connect up.

This type of light can also be wired into the supply instead of a separate shaver socket – the only difference is it connects to the *switched* live supply.

For use in bathrooms, combined light/shaver point fittings must have a pull-cord switch and an isolating transformer to BS3535.

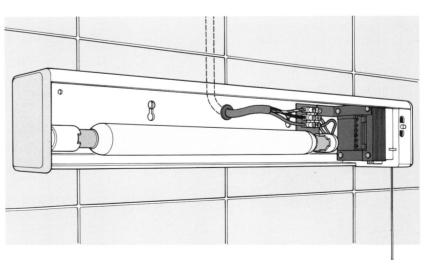

Combined light/shaver points *can easily be fitted in place of an existing wall light. Where no light already exists, you can connect the combined point like a separate shaver point except that you must wire up to the switched live terminal, not the live supply cable.*

WIRING AN EXTENSION PHONE

Since the end of 1986 you have been permitted to wire as many as three extension phones to your existing line, subject to certain limits. Most importantly, you *must* have a line with a modern, square master socket. Only British Telecom (BT) or other approved Telecomms Operator can fit these, but if you're having a phone installed for the first time, you automatically get a new-type socket. If you have an older system without one, BT will convert it for a standard charge.

All you then need is a suitable telephone and extension kit (see Shopping List). You're dealing with low voltages so there's no safety risk – and even if anything does go wrong, it shouldn't affect your existing phone.

By law, no phone equipment can be sold without an approvals sticker. You cannot connect a phone or extension kit to a BT line unless it carries the 'green circle' approved sticker. If equipment has a 'red triangle' sticker it is *illegal* to connect it to a BT line.

Trade tip

Don't overload

❝ The number of extensions you can have on one line is limited to four by the power available – overdo it, and one or more of them may not ring.

All phones, answering machines, etc have a power demand rating called a REN (Ring Equivalence Number). Ren=1 is the rating of a standard phone, but some are higher than this (a machine like a computer modem may be as high as Ren=3). If your **total** loading exceeds 4, you'll probably need another line. ❞

....Shopping List....

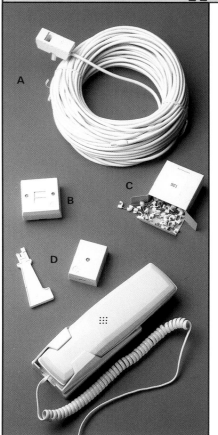

Extension kit Cable, sockets, and all the other parts you will need are available separately, but it is simplest to buy (from your nearest Telephone Shop) a complete, approved kit consisting of a plug-in converter and cable (A), extension socket (B), clips (C) and screws. Double extension packs and other kits are also available. The picture shows a BT kit; others vary in detail.

The cable supplied with the single kit is normally about 15m (16yd) long – if you need more you can buy extra and join it using a joint box (D). You will also need a joint box if you are going to install more than one socket branching from a single point (see overleaf).

Telephone Make sure that the telephone you buy has a green approval sticker (see left) and that it is suitable for use as an extension.

Tools for the job A small hammer, trimming knife, drill, screwdriver and a pair of pliers.

WHICH LAYOUT?

The first decision to make is how many sockets you want and where to put them. Even though you are limited to a total of four *phones* (a master and three extensions), you can have more *sockets* provided they aren't all in use at once. You shouldn't put sockets in areas where they will be exposed to damp or condensation such as in the bathroom, or near a sink, cooker, washbasin or toilet.

A single extension can be taken off the existing master socket (or an existing extension socket). Because the power is limited, the only restriction is that there must not be more than a *total* of 50m (55yd) of cable between the master socket and *any* extension.

Multiple extensions work on the same principle, but there are two alternative layouts. Each socket can be connected to the next one, in a chain or *series*, or they can all branch out from a single point. In the first case you can make all the connections inside the extension sockets themselves. In the second, you need a *joint box* to connect all the branches together.

A SINGLE EXTENSION

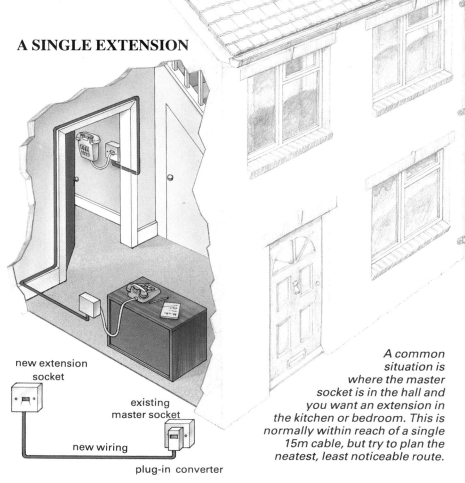

new extension socket

existing master socket

new wiring

plug-in converter

A common situation is where the master socket is in the hall and you want an extension in the kitchen or bedroom. This is normally within reach of a single 15m cable, but try to plan the neatest, least noticeable route.

RUNNING CABLE

Although few people will have the same *expertise* as a professional engineer, doing it yourself does mean that you can afford to take more *trouble* over running the wiring neatly than most professionals have the time for. Getting a neat, functional system is largely a matter of care, common sense and these simple rules:

■ Don't run within 50mm (2″) of any mains cables or sockets – they can cause interference on phone.
■ Don't run under carpets.
■ Don't run cables outdoors.

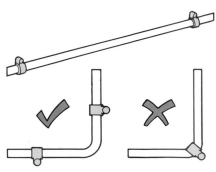

1 Clip the cable every 300mm (12in), pulling it straight as you go. Don't clip right into corners – put a clip either side, about 40mm (1½in) away.

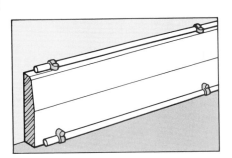

2 Clip the cable into the angle along the top or bottom of skirtings – wherever is easiest, least noticeable, and best protected from knocks.

Trade tip

Get it straight

❛ If the cable has been left coiled it can be hard to get it to lay straight. But I find that pulling it through a rag to protect my hand usually removes the kinks. ❜

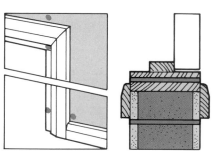

3 Don't run cable through a doorway where it may get trapped. Drill through the frame or wall at a corner – as indicated in red.

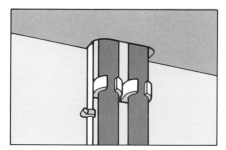

4 Going upstairs, see if you can route the cable along a cold water pipe (not a hot pipe or mains cable). Use a bent wire coat hanger to pull it through.

CONVERTING THE MASTER SOCKET

This is the property of the telephone company and must not be tampered with. Instead, you plug in a converter which has a socket for the original phone and a length of extension cable already attached.

Don't plug in the converter until you've finished the wiring: you could get an unpleasant (though not dangerous) jolt if someone rings up while you are handling the bare wires.

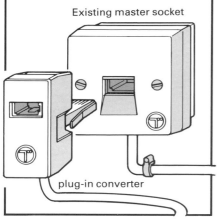

Existing master socket

plug-in converter

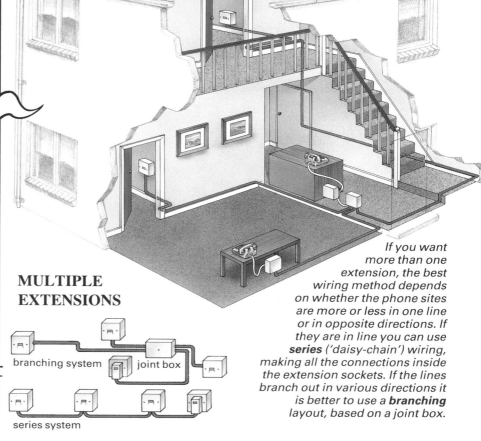

MULTIPLE EXTENSIONS

branching system   joint box

series system

*If you want more than one extension, the best wiring method depends on whether the phone sites are more or less in one line or in opposite directions. If they are in line you can use **series** ('daisy-chain') wiring, making all the connections inside the extension sockets. If the lines branch out in various directions it is better to use a **branching** layout, based on a joint box.*

STRIPPING AND CONNECTING

The outer sheath is simple to strip, as there is a drawstring to split it without damaging the conductors. There is no need to strip the inner wires, as the terminals in telephone sockets have a pair of sharp jaws which score the insulation and make their own contacts.

However, if you have not done it before, it is worth practising stripping some surplus cable. It is *not* a good idea to cut the cable to the exact length you need to start with, as a mistake means moving the socket or buying a new cable.

1 *Screw the socket to the wall and leave plenty of surplus cable beyond it. If you trim too close, there is no margin for error and no second chance.*

2 *Slit the sheath lengthways beyond the point you want to stop at (cutting around the cable is likely to nick the wires). Find the drawstring and ease it out.*

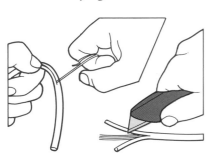

3 *Take a couple of turns of the string around your finger (if you don't it can cut you) and pull it along the cable to split the sheath. Snip off the waste.*

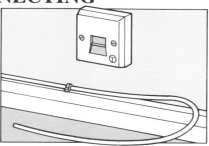

green with white ring	1
blue with white ring	2
orange with white ring	3
white with orange ring	4
white with blue ring	5
white with green ring	6

4 *There are six colour-coded conductors – three white with narrow coloured rings, and three coloured with narrow rings of white. See over for connections.*

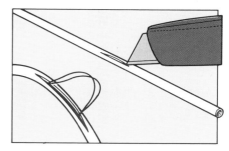

5 *Kits contain a disposable tool for inserting the wires into the terminals as shown. You must push hard to make a contact, so take care not to break the tool.*

THE RIGHT CONNECTIONS

A SINGLE EXTENSION

Trim the six wires inside the cable to leave plenty of slack and cut back the outer sheath so it will tuck neatly inside the box (use pliers to break open the notch where the cable enters the box). Connect the wires to the terminals as shown below.

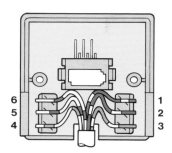

The coloured wires with white rings go to the right, the white wires with coloured rings to the left. Check that the numbers correspond to the diagram.

WIRING IN SERIES

You can use an extension socket as shown to connect two lengths of cable where you are wiring up multiple extensions in series. However, the terminals aren't big enough to take *more* than two sets of wires.

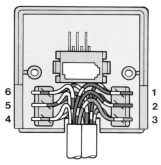

Both sets of wires are connected in the same colour order as a single extension – simply put the second set on top of the first and connect to the same terminals.

USING A JOINT BOX

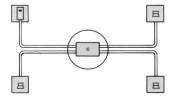

To connect more than two cables together, use a joint box. This takes one incoming and up to three outgoing cables.

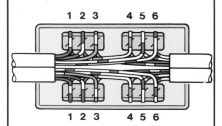

Connect the six wires in the first cable to one side of the box as shown, then connect the second set over the top. Connect the third and fourth set to the opposite side of the box in the same way.

*If using four-core cable, omit the connections to terminals 1 and 6.

■ PROBLEM SOLVER ▐

If it won't work . . .

Most likely, the wire isn't making proper contact with a terminal in the socket or joint box and needs to be inserted again. Or, a wire may have broken internally due to being kinked too often.

Neither fault is easy to see, so check back along the run to narrow down the cause. (Ideally, check each extension as you finish wiring it.)

The phone is more likely to send and receive calls than to ring, so ask a friend to call and check the ringer.

Check in this order . . .

1 Start with your old phone at the old master socket. Unplug the adaptor and check the phone still works. If it doesn't, the fault must lie in the line or old socket; get BT to check them.
2 Plug in the adaptor and try your old phone. If it doesn't

work now, the adaptor is at fault.
3 Plug your old phone into the *first* extension. If this doesn't work, the fault is in the cable from the adaptor, or its socket connection.
4 & 5 Work down any other extensions in sequence away from the first. If there is a fault, it must lie in the area shown shaded red in the diagram.

Finally, test your new phone in a socket you know works.

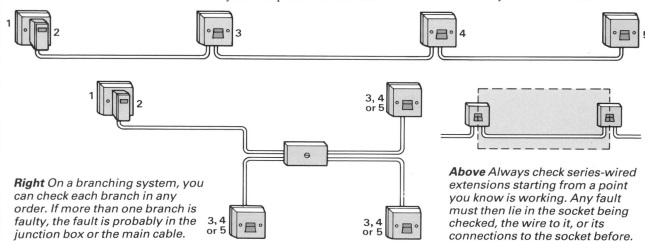

Right On a branching system, you can check each branch in any order. If more than one branch is faulty, the fault is probably in the junction box or the main cable.

Above Always check series-wired extensions starting from a point you know is working. Any fault must then lie in the socket being checked, the wire to it, or its connections to the socket before.

ADDING EXTRA TV AERIAL SOCKETS

In most reception areas, a television set gives a much better picture with a fixed roof aerial. Usually, there's only one room with an aerial connection, and to watch TV anywhere else you either need to use a portable aerial or plug in a clumsy aerial extension lead. But in fact it's neither difficult nor expensive to connect several sockets to the roof aerial. This allows you to:

■ Plug a TV into the roof aerial anywhere in the house.
■ Use another TV at the same time.
■ Connect up your video so that you can play it back through a TV set in another room.

Even if you're not keen on tackling mains electrical work, aerial wiring is one job you can do with complete confidence. There is no risk of shocks and no risk of damage to the set. The only limitations on aerial extensions are:

■ You must NOT take a mains powered TV into the bathroom – this is extremely dangerous.
■ In poor reception areas, or if the cables are very long, you may need to fit a device which boosts the signal. This is covered in detail in Problem Solver.

A second TV in the bedroom or elsewhere in the house can have just as good a picture as the main set if you install a branch cable off the roof aerial.

.... Shopping List ....

The main fittings for extending aerial wiring are cable, plugs, sockets and signal splitters.

Aerial cable TV aerials are wired up with 'low-loss' UHF *co-axial* cable (sometimes called 'co-ax'). This usually has brown or white outer insulation. Other types of co-ax are made for audio wiring, so be sure to use the right sort.

Plugs and sockets Co-ax cable is connected up with co-axial plugs and sockets. The aerial cable connects to the set via a 'male' co-axial plug. 'Female' plugs are used on video connections and for extending cables fitted with male plugs.

For a neater installation, you can fit wall-mounted sockets and connect the TV via a short length of cable (known as a fly lead) with a plug at each end.

Signal splitters enable you to divide the aerial signal between two or more TV sets. There are several types (see overleaf).

Signal boosters or filters may be needed to solve reception problems. These are covered in detail in Problem Solver.

Tools: Screwdrivers, electric drill and bits, trimming knife, hammer. A cheap pair of wire strippers is worth having, but you can manage with a knife.

Aerial fittings (right) are widely available from hobby electronics shops and DIY stores.

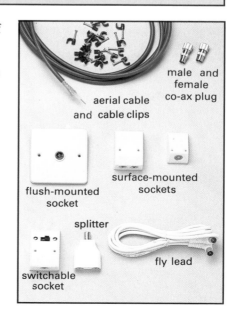

aerial cable and cable clips

male and female co-ax plug

flush-mounted socket

surface-mounted sockets

switchable socket

splitter

fly lead

EXTENDING YOUR SYSTEM

A typical roof aerial has a length of co-axial cable running down the side of the house. This usually comes in through a window frame and plugs into the back of the set. Or it may lead to a socket mounted on the skirting board or wall, with a separate length run to the TV.

Alternatively, the cable may enter the house via the loft, or the aerial itself may be installed in the roof space. This makes little practical difference, except that the conversion is simpler because all the cables are accessible from indoors.

What you need to add
To extend the system, you break into the cable coming from the aerial at a convenient point, and install a signal splitter and branch cable. Where to do this depends on:

Convenience The shorter the cables, the better. And think about the ease of running the cable from the branch point to the TV.

Accessibility There's no point in trying to branch off at a point you can't get at easily.

Cover Don't branch the cable at any point where it will be exposed to the elements. If you can't avoid this, use a weatherproof splitter.

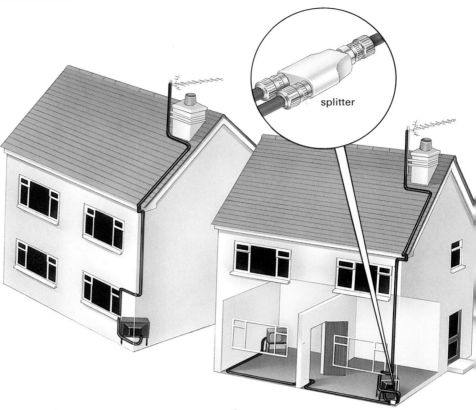

splitter

The lead from the roof aerial (above) *commonly runs down the outside of the house and enters via a hole in the living room window frame.*

One way to extend the system (above) *is to fit a signal splitter to the end of the existing cable and run a new length to the second socket.*

WORKING WITH CO-AXIAL CABLE

Co-ax cable is the heart of an aerial system. The single insulated core in the middle, which carries the signal, is surrounded by a braided sheathing of fine wire strands under the outer plastic insulation. This stops interference reaching the core.

It's worth practising stripping and connecting a spare piece of co-ax. The main thing to ensure is that none of the fine strands of the sheathing touch the inner core or its terminal. There is no risk of a 'short circuit' – it simply means that the TV signal could be interrupted.

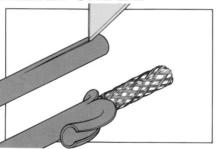

1 Strip co-ax by slitting the sheathing for 25mm (1″) or so with a trimming knife. Cut off the insulation and fold back the sheathing to expose the core.

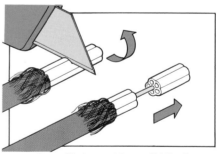

2 Cut off about 12mm (½″) of the inner insulation to expose the core wire. This insulation is often cellular, and needs care to cut it cleanly.

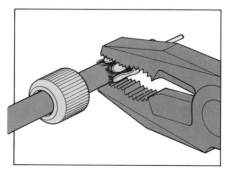

To fit a plug, *unscrew the cap. Slip this and the sheathing grip over the cable so the grip encloses the braid completely, then close the grip with pliers . . .*

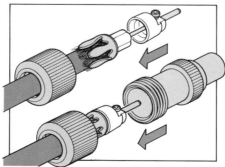

. . . **Slip the core** *wire into the centre pin, which may have a clamp screw. Cut off any surplus wire projecting from the end. Then reassemble the body.*

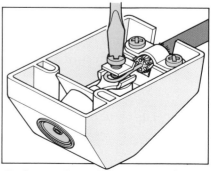

To fit a socket *wind the core wire clockwise round the inner terminal screw and tighten. Trap the sheathing in the outer clamp. Make sure no stray strands touch.*

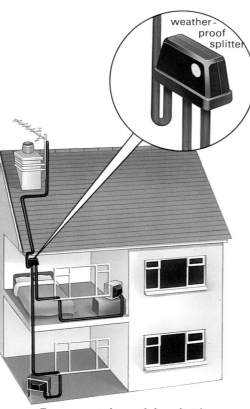

For an upstairs aerial socket it may be better to fit the signal splitter partway down the cable, nearer to the TV, to keep the leads short.

TYPES OF SIGNAL SPLITTER

Splitters divide a single aerial signal to feed two or more TVs. Some types are combined with signal boosters for use in poor reception areas (see Problem Solver).

Plug-in splitters are the simplest type. They fit into plugs attached to the ends of the aerial leads.

Surface-mounting splitters combine a socket for the first TV with a connection for the second.

Switchable sockets allow you to divert the aerial signal from one outlet to another.

Weatherproof splitters can be installed outdoors – usually on the aerial mast – with leads to feed more than one socket.

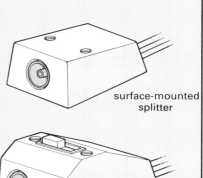

surface-mounted splitter

switchable socket

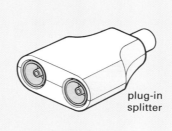

plug-in splitter

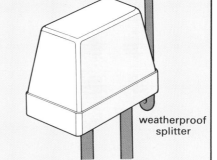

weatherproof splitter

RUNNING THE CABLES

Aerial cable is bulky, and won't run round tight corners easily. For this reason, route it inconspicuously wherever possible, and don't pull it tight or force it around bends.

Run the cable under the floor-boards if these are accessible and you can lift them easily. Otherwise take it along the tops of skirting boards, either clipped in place or hidden in plastic mini-trunking.

Don't be tempted to run the cable under a carpet where people can walk on it – this is likely to damage both, the cable and the carpet. Where it is exposed, decide whether white or brown sheathing is least obvious.

Sockets come in different styles and can be bought with single or double outlets or combined with splitters (see above). The main choice is between surface mounting or flush mounting fittings.

Flush mounted fittings are exactly the same size as power sockets and are designed for use with standard socket or switch backing boxes, for fitting to the wall itself.

Surface mounted fittings need no backing box and normally screw to the skirting board.

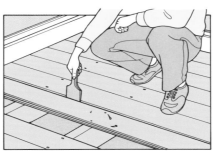

Under floorboards is one route for the cable, if you can get at the boards easily. Use a bolster, spade or crowbar to lever up a series of boards for access.

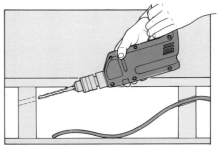

Drill holes in the joists to feed the cable through and let it hang slackly between them. Drill a hole in the floor to feed the cable up to the skirting.

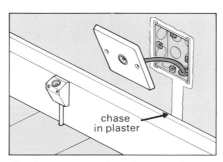

Surface mounted sockets can be screwed to the skirting. To fit a flush wall socket, you need to cut a channel in the plaster and fit a backing box to the wall.

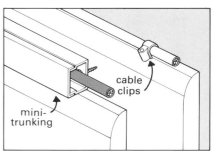

Run exposed cable along the top of the skirting board with cable clips. Alternatively route it in plastic mini-trunking, nailed or glued to the wall or skirting.

VIDEO PLAYBACK

It is possible to play back a video recorder over your aerial system so that you can view it on any TV in the house. This needn't prevent you from watching transmitted TV at the same time – the only disadvantage is that the video won't be controllable from the extension.

The cheapest option is to connect the video so that it receives a signal straight from the main aerial. Use a splitter at this point to distribute the video output to both the main TV and the extension. You can then tune either TV to a transmitted channel or select the video channel to view a playback.

This arrangement works, but it is possible that the picture quality may suffer because of the long leads and multiple connections. An alternative is to use a *distribution amplifier* designed to work with video. These are powered from the mains and cost around as much as three or four video cassettes. They have a single aerial input and two or more amplified outputs to compensate for the loss of signal strength.

You can also buy *video sender units* which plug into your video recorder and transmit signals to all the TVs in the house. However, while it is legal to sell them in the UK, it is illegal (for reasons of interference) to use them!

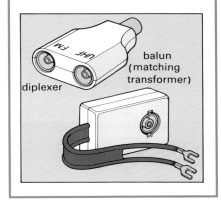

Connect your video like this *as an inexpensive means of playing it back over a TV plugged into an extension socket, as well as the main TV. Select either the video or transmitted signals in the ordinary way.*

video out — aerial in

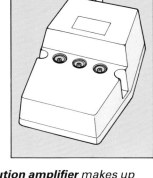

A distribution amplifier *makes up for the loss in signal strength caused by splitting the signal from the video in the way shown on the left. It needs a mains power supply and may have a variety of different outputs.*

PROBLEM SOLVER

Signal strength problems

It is important for the signal reaching the set to be the right strength, otherwise the picture goes fuzzy and won't hold. Colour TVs need a stronger signal than black and white, and tend to suffer more.

Extending the aerial causes a slight loss of strength, and if the picture gets worse after you have completed the job, you may need to fit a booster.

Signal boosters

Indoor signal boosters and *mast-head amplifiers* (fitted at the aerial) are mains-powered. They compensate for loss of signal strength caused by a long cable run or by multiple extensions. However, the booster may also increase the strength of any interference present – though

you can get devices called *filters* designed to remove this. Many boosters have multiple outlets for a number of different TV sets, so you can do without separate splitters.

Aerial alignment

If the picture is unsatisfactory to start with, make sure that the aerial is properly aligned. This isn't all that easy, and if the aerial is hard to reach you may prefer to call in an aerial installer.

Professional installers use special *field strength meters* to set the aerial up. But if you want to have a go yourself and can get to the aerial, do the job when the test card picture is being broadcast. Get someone to watch the set and call out whether the picture gets better or worse as you move the aerial.

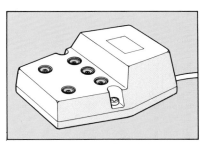

An indoor booster *fitted in the main cable increases the signal strength to extra outlets.*

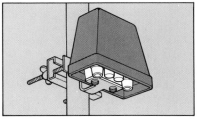

A mast-head amplifier *has a weatherproof casing and may have several outlets.*

TWO-WAY LIGHT SWITCHING

Most fixed lights in the home are controlled from single switches, an arrangement known as *one-way switching*. However, there are several places where it is more convenient if you can turn the same light on and off from two positions – a set-up known as *two-way switching*.

The most commonly two-way switched light is the one on the landing, which for safety and convenience is usually controllable both from a landing light switch and from another in the hall. Other popular locations for two-way switching are a large living room with separate entrances, and a bedroom with wall lights for reading. In both cases, it's helpful to be able to control the same light(s) from more than one place. It is even possible to provide switching from three or more positions if required, by using an extension of the two-way switching principle. A typical use would be for switching hall and landing lights from half landings.

THE BASIC CIRCUIT

For one-way switching, a standard *two-core and earth* cable links the light fitting (loop-in wiring) or its junction box (junction box wiring) to the switch controlling it. The red and black cores, which are both live, connect to the top and bottom terminals on a *one-way switch*, or to the terminals marked C and L2 on a *two-way switch* if this is wired for one-way operation (the L1 terminal is left unused; on some makes of switch the terminal labelling is slightly different – see below).

For two-way switching, *a two-way switch* is fitted at both switch positions. The two-core cable from the light fitting or junction box runs to the first switch, where the red goes to L1 and the black to L2. This switch is then linked to the second via a length of *three-core and earth* cable (called *strapping* cable), where the cores are connected as shown on the right. The rest of the light circuit remains unaffected, but the light can now be controlled from either switch.

Two-way or Two-gang?

Don't confuse *two-way* with the term *two-gang*: a *two-gang* switch faceplate holds two switches and controls two lights separately, though most two-gang switches are also two-way. (You can in fact have up to six gangs on one switch plate).

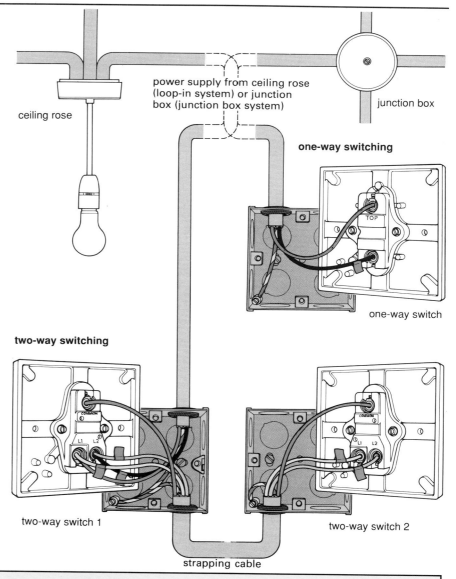

power supply from ceiling rose (loop-in system) or junction box (junction box system)

ceiling rose

junction box

one-way switching

one-way switch

two-way switching

two-way switch 1

two-way switch 2

strapping cable

CORE CONNECTIONS

The cores in three-core and earth cable are colour-coded red, blue and yellow. This is purely for identification purposes; all are live when the light is on, and the blue and yellow cores should be 'flagged' with red PVC tape or sleeving to show this.

It doesn't matter where the cores go so long as like cores link like terminals. However, it is accepted practice to link the top C terminals in two-way switches via the red core, the L1 terminals via the yellow core and the L2 terminals via the blue core.

Terminal markings
Some makes of two-way switch have their terminals labelled slightly differently from those described in this section; wire to the corresponding terminals.
C may be labelled L1 or 1
L1 may be labelled L2 or 2
L2 may be labelled L3 or 3

CONVERTING TO TWO-WAY SWITCHING

Converting a one-way switching arrangement to two-way switching is normally a simple matter of mounting a new two-way switch in the desired position, then linking it to the existing switch via a length of three-core and earth cable. There may not even be any need to change the existing switch (see Tip).

Before shopping for parts, decide how to mount the new switch and where to run the strapping cable.

The switch can be flush mounted on a metal backing box buried in the plaster (in which case the cable should be buried too), or surface mounted on a plastic pattress box with the cable clipped to the wall or run in plastic mini-trunking. Yet another alternative is to use a *5 amp*

two-way ceiling pull-cord switch so that you don't have to run the cable down the wall to it.

The strapping cable can normally run up the wall from the existing switch, through the upstairs ceiling void, and down to the new switch. Where this isn't possible, an alternative route is down the wall and along the top of the skirting.

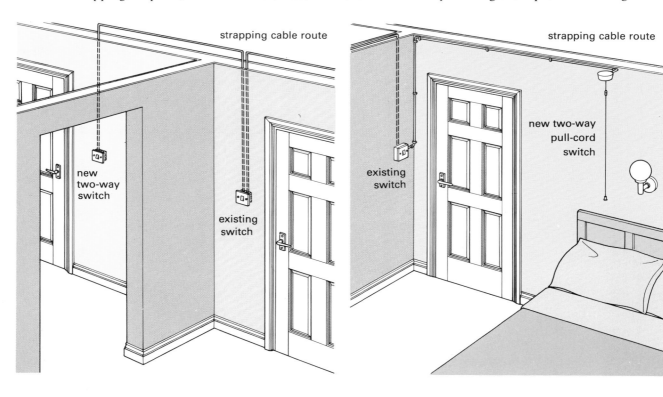

Option 1 – the strapping cable runs up through a chase in the plaster from the existing switch, through the ceiling void, and down through a similar chase to the new two-way switch position.

Option 2 – from an existing surface mounted switch, the cable is clipped along the edge of the wall as far as a new pull-cord switch. A neater layout can be considered if and when the room is redecorated.

.... Shopping List ....

For a simple two-way conversion you need one or two new *two-way switches* (see right), suitable *mounting box(es)*, and *1.0mm² three-core and earth PVC-sheathed cable* to link the old and the new switches. Sundry materials include *wall fixings*, *red PVC sleeving or tape* to mark the switch wires, and *green/yellow PVC sleeving* to cover the bare earth cores.

For parts for other conversions, see overleaf.

For surface mounting you also need *cable clips* or *plastic mini-trunking.*

For flush mounting, you need materials for making good the wall.

Tools checklist: Wire strippers, electric drill and bits, screwdrivers, hammer and bolster, tools for making good (maybe).

Trade tip

Which switch?

❝Before buying any parts for the conversion, turn off the mains supply at the consumer unit and unscrew the existing switch from its mounting box. If it is a one-way type, you must replace it with a two-way switch (though this can go in the same backing box). However, it's not unusual to find two-way switches wired for one-way operation, in which case you can use the existing switch.

Multi-gang switches are nearly always two-way, but check – there should be three terminals for each gang. In practice you may need to replace the existing box with a deeper one to accommodate the extra strapping cable (see Problem Solver).❞

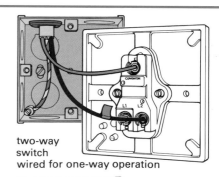

two-way switch wired for one-way operation

two-gang switch (both gangs wired for one-way operation)

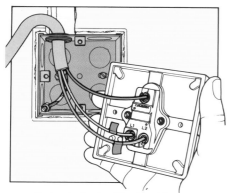

1 Cut a hole for the backing box holding the new switch (or screw the box to the wall). Run in and connect the strapping cable, not forgetting the earth.

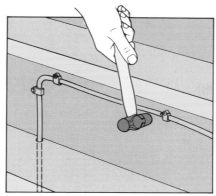

2 Run the strapping cable back from the new switch towards the existing switch position – up the wall and through the ceiling void is the usual route.

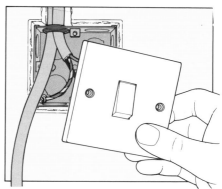

3 Turn off the mains, open up the existing switch, then run in the strapping cable – you may be able to follow the route of the existing switch cable.

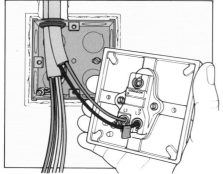

4 *If the switch is one-way,* disconnect the existing cable cores and fit the new switch. Reconnect the colour cores to L1 and L2, and the earth to the box.

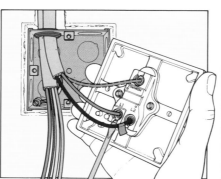

5 *If the switch is two-way,* simply transfer the red core from its present position on C to L1. Leave the black and sleeved earth cores as they are.

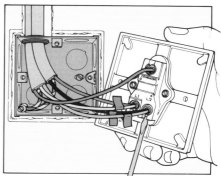

6 Connect the strapping cable cores to the appropriate terminals on the switch: red to C, yellow to L1, and blue to L2. Connect the earth to the box.

▐ PROBLEM SOLVER

No room for the cable

Most switches are fitted to shallow plaster-depth mounting boxes. These don't have much space inside, and to make room for the extra strapping cable cores – especially when converting multi-gang switches – you may need to replace the existing box with a deeper one of the same width. The job involves disconnecting all the existing cores from the switch, so make sure you label them first.

■ Turn off the mains supply and release the faceplate from the mounting box. Using self-adhesive labels, mark the back of each gang 1, 2 and so on. Then label each of the cores with the switch number followed by the terminal marking to which it is connected. So, for example, you would label the core which is connected to terminal L1 of switch 2 "2/L1". (If the terminals are marked 1, 2

and 3, use letters A, B etc. for the gangs to avoid any possible confusion.)

■ Disconnect the cores from the switch and the earth wires from the mounting box. Remove the screws securing the mounting box, and prise it out of its recess.

■ Enlarge the depth of the recess to take the new box. The simplest way to do this is by drilling a series of holes in the back to the right depth, using a masonry bit. Then chip out the debris with a hammer and cold chisel. Afterwards, drill and plug new fixing holes, then screw the box to the wall and make good around the recess with filler or repair plaster.

■ Reconnect the labelled cores to the switch terminals and the sleeved earth wires to the earth terminal in the mounting box.

■ Complete any new wiring, then refit the switch faceplate.

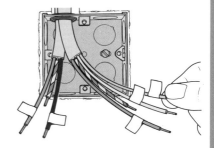

Label the cores of the cables to indicate which terminal they were connected to. Then disconnect them and remove the switch.

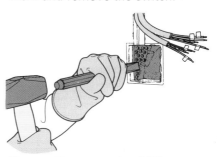

Deepen the recess by drilling a series of holes to the same depth, then chop out the waste. Fit the new box and make good.

MULTI-GANG CONVERSIONS

Multi-gang switches may appear to complicate matters, but as far as two-way switching is concerned the principle remains the same.

A common arrangement is to have a two-gang two-way switch in the hall, with one gang controlling a one-way switched hall light and the other gang providing two-way switching for a landing light. In this case a worthwhile improvement is to convert the landing light switch from a one-gang two-way switch to a two-gang two-way switch as well. This allows the downstairs hall light to be controlled from upstairs in the same way as the upstairs light is controlled from downstairs.

Having fitted the new switch, all you have to do is run a new strapping cable from the gang which you select to control the hall light to the gang on the downstairs switch currently controlling it. Make the cable connections as shown in the diagram below right.

If the existing strapping cable was run in conduit, you may be able to feed the new cable through the same way. Otherwise, surface-mount it for now and chase it in alongside the existing switch cable the next time your redecorate.

When chasing in, take care not to damage the existing cable – it may not run vertically between ceiling and switch. It's safer to turn off the power **at the mains** (the hall switch may be linked to both the upstairs and downstairs lighting circuits), then expose the cable route at ceiling level so that you can mark its likely route down to the switch.

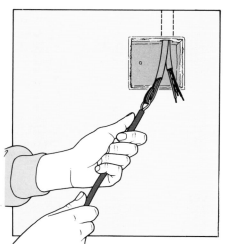

Where the existing switch cable is chased in to the plaster with conduit, you may be able to draw the new cable through it as well, using stiff wire or an electrician's draw tape.

ORIGINAL ARRANGEMENT

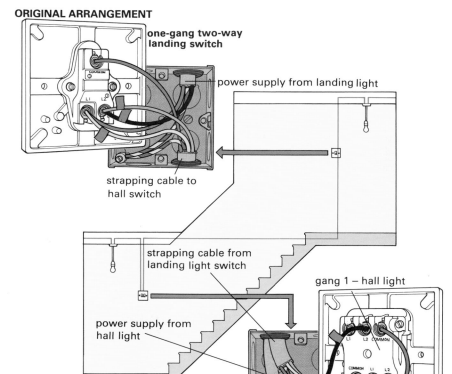

one-gang two-way landing switch

power supply from landing light

strapping cable to hall switch

strapping cable from landing light switch

power supply from hall light

two-gang two-way hall switch

gang 1 – hall light

gang 2 – landing light

NEW ARRANGEMENT

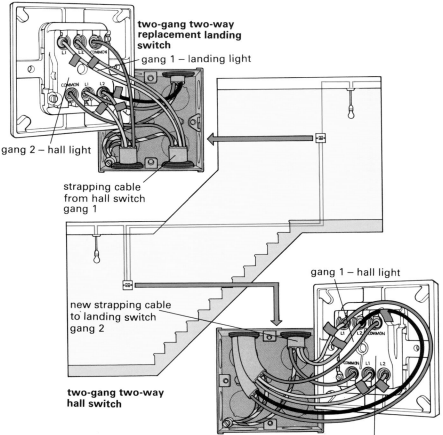

two-gang two-way replacement landing switch

gang 1 – landing light

gang 2 – hall light

strapping cable from hall switch gang 1

new strapping cable to landing switch gang 2

gang 1 – hall light

two-gang two-way hall switch

gang 2 – landing light

THREE-WAY SWITCHING

The two-way switching principle can be extended to provide three-way (or more) switching by using a special 4-terminal *intermediate* switch. Three-way switching is usually enough for most homes, but a large staircase with several landings may benefit from a switch at each level.

For three-way switching, the two-way circuit is connected to the ends of a strapping cable in the usual way. Between them is connected the intermediate switch; the blue and yellow cores on the cable from the 'master' switch (the one containing the power cable) are then wired to the upper pair of terminals, while those from the second two-way switch are connected to the lower pair.

The red cores on the strapping cable are not connected to the intermediate switch, but are linked directly to each other with a *5 amp terminal block connector* which is then folded into the mounting box. As usual, all earth cores are sleeved and connected to the terminal on the box itself.

Further intermediate switches can be added in exactly the same way.

Two-way switching in both directions between the hall and landing can be a major safety improvement.

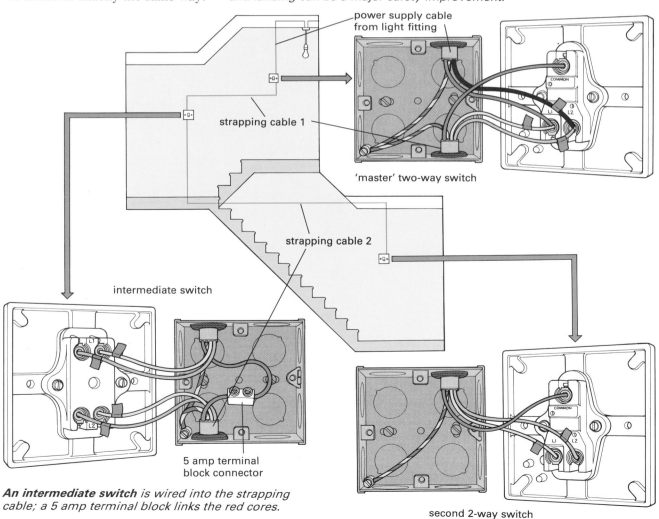

power supply cable from light fitting

strapping cable 1

'master' two-way switch

strapping cable 2

intermediate switch

5 amp terminal block connector

An intermediate switch is wired into the strapping cable; a 5 amp terminal block links the red cores.

second 2-way switch

SWITCHING FOR WALL LIGHTS

In living rooms and bedrooms, it's often convenient to have more than one light under two-way switch control. A typical set-up is to have the room's main central light on one-way control, switched from near the door, and to provide two-way switching for the wall lights – from near the door, and at some other point in the room such as beside the bed in a bedroom.

The simplest way of arranging this is to wire all the switching cables – plus the power cable from the main lighting circuit – via a multi-terminal junction box mounted in the ceiling void. Make the connections to the box as shown in the diagram, then simply run the cables through the ceiling void and down the wall to the appropriate switches.

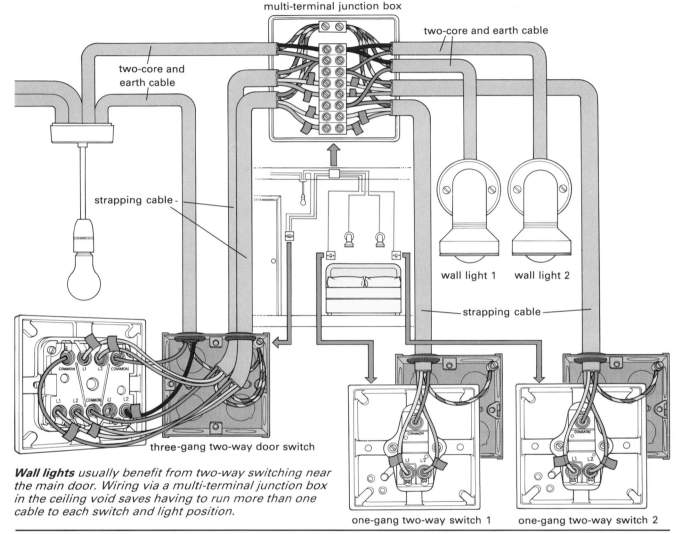

multi-terminal junction box

two-core and earth cable

two-core and earth cable

strapping cable

strapping cable

wall light 1   wall light 2

three-gang two-way door switch

one-gang two-way switch 1   one-gang two-way switch 2

Wall lights usually benefit from two-way switching near the main door. Wiring via a multi-terminal junction box in the ceiling void saves having to run more than one cable to each switch and light position.

TWO-WAY DIMMER SWITCHING

You can combine two-way switching of one or more lights with dimmer control by intalling a dimmer switch in place of the original one-way switch and using a standard two-way switch at the new position. Alternatively, some types of dimmers can be fitted with extension controls instead, to allow dimming from both points. And in either case you can add intermediate switches between the two if required.

Where the dimmer has standard C, L1 and L2 terminals, the strapping cable connections are the same as for ordinary two-way switching. However some dimmers have their own special terminal markings, in which case be sure to follow the maker's own wiring diagrams.

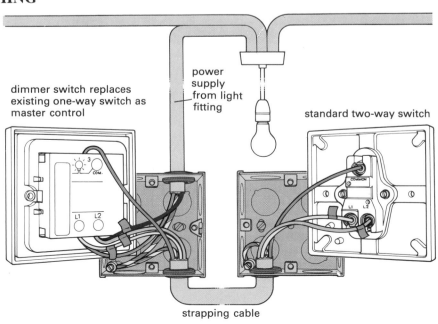

dimmer switch replaces existing one-way switch as master control

power supply from light fitting

standard two-way switch

strapping cable

FITTING A NEW PENDANT LIGHT

A simple pendant bulb holder tends to cast a harsh, shadowy glare over a room, even when heavily shaded. Spots, tracks and wall lights are all good alternatives, but fitting any of these is likely to involve running new wiring.

A simpler solution is to replace the existing pendant with a modern adjustable or glare-free design such as the one shown below. You may also want to take the opportunity to move the light to a better position – centred over the table for example.

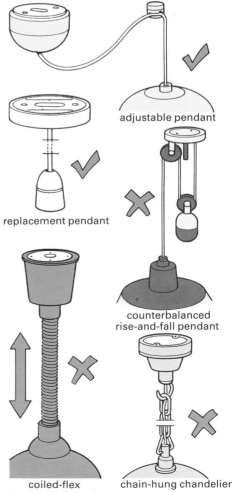

adjustable pendant

replacement pendant

counterbalanced
rise-and-fall pendant

coiled-flex
rise-and-fall pendant

chain-hung chandelier

Trade tip

Check the rose fixings

❝ If the existing ceiling rose is loosely fixed – as they often are – fitting a new pendant light of any type is likely to make matters worse.

You should be able to check without removing the cover. If there's any sign of movement, make sure you have some plugging compound and hollow wall fixings before you start. ❞

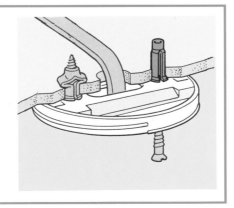

FITTING THE NEW LIGHT

Turn off the power at the consumer unit and remove the appropriate lighting circuit fuse (or trip the appropriate MCB). Switch the light on and off to double check.

If, as is most likely, the new light has a plastic bulb holder and body, the flex will have two cores – brown for live, blue for neutral. But lights with metal fittings may have a three core flex incorporating a green/yellow insulated earth wire.

No matter how the rose is wired (see right), you simply connect the new flex cores in place of the old ones. But note the following:

■ If the flex cores are not coloured (two-core flex only), it doesn't matter which way round they go.

■ If the original flex is very old and has the old style colour coding, connect brown in place of red and blue in place of black.

■ If the old flex has two cores, but the new one has three, the extra green/yellow flex core should be connected to the terminal occupied by the green/yellow (or unsheathed copper) circuit cable cores ('Earth' in the diagrams).

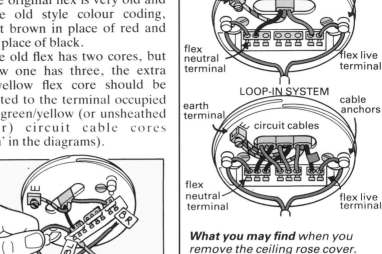

What you may find when you remove the ceiling rose cover. Older-style junction box wiring (top) leaves one circuit cable visible at the rose; new-style loop-in wiring (above) leaves two or three. The type of wiring employed does not affect how you connect the new flex.

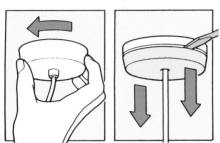

1 With the power off, remove the ceiling rose cover. Some types simply unscrew; others are a push-fit and can be prised off with a screwdriver.

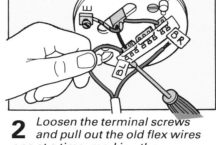

2 Loosen the terminal screws and pull out the old flex wires one at a time, marking the relevant terminals with pieces of tape as you go.

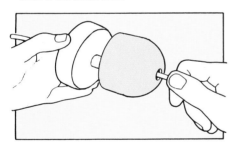

3 Cut the new flex if you need to alter the height of the light. Feed the flex through the hole in any trim cover supplied, then through the rose cover.

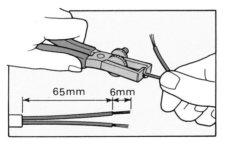

4 If the end of the new flex isn't prepared, strip back the outer sheath for about 65mm (2½"), then bare the ends of the cores about 6mm (¼").

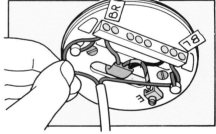

5 Connect the wires to the appropriate terminals and loop them around the anchors in the rose. Refit the rose cover, then push on the trim cover (if fitted).

PROBLEM SOLVER

Loose roses

A properly fitted ceiling rose will be screwed to a piece of board nailed between the upstairs floor joists or direct to a joist. These rarely come loose.

However, it's not uncommon for the rose to be part-screwed to a joist, to the laths in an old lath and plaster ceiling, or straight through the plasterboard in a new ceiling. Any of these 'bodges' will cause the screws to come loose as soon as any weight is put on them.

The answer is to screw into something more secure. Plastic wing cavity wall fixings are particularly good for plugging

holes made into plasterboard, while spring toggle cavity fixings help to spread the load in a lath and plaster ceiling. But if the screw has pulled out a chunk of

plaster, patch the hole with plugging compound and screw into this.

Be sure to turn the power off before working on the rose.

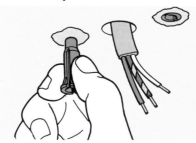

Plug loose rose fixing holes with hollow wall fixings or a similar lightweight cavity fixing.

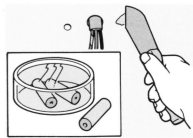

Use plugging compound if the screws have torn away some of the ceiling plaster.

FITTING WALL LIGHTS

Wall lights can provide soft background lighting, highlighting for pictures or displays of ornaments, or localized task lighting – such as reading lights above a bed. There are dozens of styles to choose from, and several ways to wire them up.

Planning the job

Once you have decided what lights to install and roughly where to put them, your main task is to plan the wiring. This means thinking about how to provide a power supply to the lights, and how to switch them on and off. The panel below lists the options, which are detailed overleaf.

The other problem you face is concealing the wiring. Unless you are prepared to put up with cable runs on wall surfaces or within mini-trunking, you will have to cut chases in the plaster (or access holes in stud walls).

This inevitably means ruining existing decorations. If you are not planning to redecorate in the near future, a fair compromise is to clip the cables to the wall and disguise them with paint, then conceal them properly when the room is eventually redecorated. Remember to leave some slack if you take this option.

Wall lights *have become increasingly popular as a means of complementing or even replacing traditional pendants.*

POWER SUPPLY AND SWITCHING OPTIONS

Your supply options are:
■ To connect to an existing lighting circuit, at a ceiling rose or junction box, or by wiring a junction box into the circuit cable at a convenient point.
■ To connect to a fused spur wired from a ring or radial power circuit.

If you opt to extend a lighting circuit, check that your new lights won't overload it. Each circuit can supply a maximum of 12 fittings (each counts as a nominal 100W, even if the bulbs used are actually smaller than this), so count up how many the circuit already supplies before proceeding.

As far as switching is concerned, lights wired from a lighting circuit can be:
■ Switched in tandem with the existing pendant light(s).
■ Switched in place of the

pendant light(s) using the existing wiring.
■ Switched independently using new switches and switch wiring.

Lights wired from a power circuit can be switched from the fused connection unit supplying them, or via a new plateswitch.

Although the switching method is partly a matter of personal preference, it's sensible to choose whichever wiring layout involves the least disruption. If you opt for tandem or replacement switching, you must be able to gain access, from above, to the existing ceiling roses and junction boxes. Independent switching involves more new wiring, but also gives you more flexibility.

In all cases, turn off the power at the consumer unit and remove the circuit fuse before starting work on the wiring.

....Shopping List....

What cable and wiring accessories you need in addition to the lights themselves depends entirely on how you arrange the wiring. Read through the options described overleaf, decide which suits you best, then make up a list accordingly.
Sundry parts not detailed overleaf include cable clips, green and yellow PVC earth sleeving, and red sheathing for marking the switch live cores. For enclosed light fittings you need heat-resistant sleeving to protect the wiring within the fitting (this may be supplied with the fitting).
Other materials include filler or repair plaster for making good, and wall fixings or machine screws for securing the lights (see page 68).
Tools checklist: Screwdrivers, wire strippers, electric drill and bits, club hammer and bolster, tools for lifting floorboards, tools for plastering/making good.

WALL LIGHT WIRING ARRANGEMENTS

OPTION 1: TANDEM SWITCHING

If you want your wall lights to come on at the same time as the main room light, and to be controlled by the same switch, you need to take a spur cable from the supply to the existing light.

On a loop-in system, connect the spur cable cores to the light's loop-in rose as shown.

On a junction box system, you can either wire the spur cable from the existing rose, or from the four-terminal junction box supplying it – whichever is more accessible.

Run the new spur cable from here to the new light fitting and connect using one of the methods shown on page 68.

Multiple lights

Where the spur is to feed two or more lights, run the spur cable to a three-terminal junction box close to their planned positions. Then run separate branch cables from this box to each light in turn as shown.

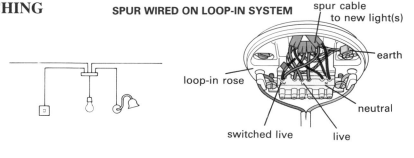

SPUR WIRED ON LOOP-IN SYSTEM

spur cable to new light(s)
earth
loop-in rose
neutral
switched live
live

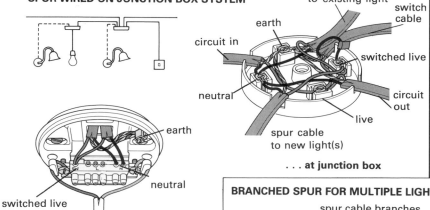

SPUR WIRED ON JUNCTION BOX SYSTEM

to existing light
switch cable
earth
circuit in
switched live
neutral
circuit out
live
spur cable to new light(s)

. . . at junction box

earth
neutral
switched live

. . . at ceiling rose

BRANCHED SPUR FOR MULTIPLE LIGHTS

spur cable branches (to new lights)
spur cable

3-terminal junction box

OPTION 2: REPLACEMENT SWITCHING

If the wall lights are to take the place of the existing pendant light and you intend to remove the latter altogether, you can use its wiring to supply the new lights with only minimal disruption.

■ Disconnect the wiring to the existing light. If there are several cables (and so a loop-in system), label which is which.

■ Either remove the rose and fitting then make good the ceiling, or just remove the flex and lampholder and leave the rose for possible future use.

■ Working from above, reconnect the cable(s) to a junction box screwed to a nearby joist; use a 3-terminal box if there's only one cable, a 4-terminal box for more.

■ Run a spur cable from here to the new light fitting – or, for multiple lights, via another 3-terminal junction box as for Option 1. The new light(s) will be henceforth controlled by the existing light switch, with no need for any alterations to the switching arrangements.

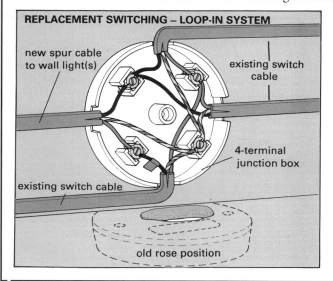

REPLACEMENT SWITCHING – LOOP-IN SYSTEM

new spur cable to wall light(s)
existing switch cable
4-terminal junction box
existing switch cable
old rose position

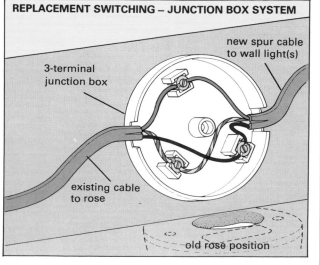

REPLACEMENT SWITCHING – JUNCTION BOX SYSTEM

new spur cable to wall light(s)
3-terminal junction box
existing cable to rose
old rose position

OPTION 3: INDEPENDENT SWITCHING

Adding wall lights with separate switching is a little more complex. Assuming that you can take your power supply from an existing lighting circuit without overloading it, your first step is to decide on a convenient connection point. Make sure this is actually on the main circuit, and not on a switch cable or a spur to an existing light.

■ Cut the circuit cable and insert a 4-terminal junction box.

■ Reconnect the main circuit cores as shown, then connect in a spur cable to run to the new light(s), and a switch cable to run to a convenient switching position.

■ Use a 3-terminal junction box as for Options 1 and 2 to split the supply to several lights.

■ Run the cable to the new switch as shown in the panel on the right.

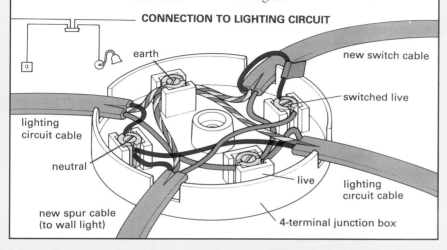

CONNECTION TO LIGHTING CIRCUIT

earth

new switch cable

switched live

lighting circuit cable

neutral

live

lighting circuit cable

new spur cable (to wall light)

4-terminal junction box

OPTION 4: WIRING FROM A POWER CIRCUIT

If it is more convenient to take the supply from a power circuit, start by selecting a suitable connection point. This could be an existing unspurred socket, or a 30 amp junction box wired into the main circuit cable (the rules are the same as those for adding an extra socket).

■ From here, run a spur in 2.5mm² two-core-and-earth cable to a fused connection unit (FCU) with cable outlet near the new wall light(s).

■ Fit a switched FCU if you want to control the lights from this. Then run 1.0mm² cable from the FCU direct to the light, or via a 3-terminal junction box for multiple lights.

■ If you want conventional switching using a plateswitch, fit an unswitched FCU. Run 1.0mm² cable from here to a 4-terminal junction box, then run cables from the box to the switch and light(s) as described in Option 3 and the panel (right).

Fit a 5-amp fuse in the fused connection unit.

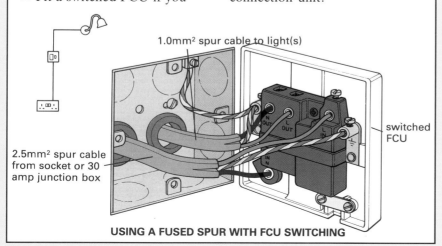

1.0mm² spur cable to light(s)

2.5mm² spur cable from socket or 30 amp junction box

switched FCU

USING A FUSED SPUR WITH FCU SWITCHING

SEPARATE SWITCHING

Options 3 and 4 both provide for separate switching via a conventional plateswitch.

■ If you want the wall light switch next to the existing room light switch, you may be able to feed the new switch cable from the 4-terminal junction box down to this switch position through existing conduit.

If you can, replace the existing single-gang switch with a two-gang one, and connect the existing and new switch cables to their respective terminals so one gang controls the room light and the other the wall light(s).

On a flush mounted fitting, bear in mind that you may have to exchange the backing box for a deeper one to make room for the extra cable.

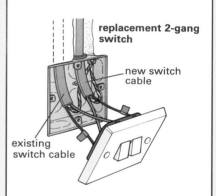

replacement 2-gang switch

new switch cable

existing switch cable

■ If you can't find a way to feed the new switch cable alongside the existing one, the switch can go anywhere. Even so, a position close to the existing pendant switch will probably be most convenient.

For this option, simply cut a new chase and hole for the backing box, then wire the cable to a single-gang plateswitch as shown.

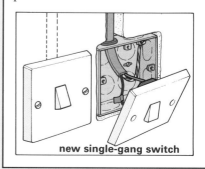

new single-gang switch

FITTING THE LIGHTS

The new light fittings may have either an *enclosed* or an *open base*, or be fitted with the recently introduced *LSC (luminaire support coupler) plug* to allow for easy removal.

Enclosed base fittings can be screwed directly to the wall. The spur cable cores connect to terminals within the base.

Open-base fittings must be mounted over an approved heat-resistant enclosure recessed into the wall; the spacing of the mounting holes determines which type you use:

■ For holes spaced 51mm (2") apart, use a *circular conduit box* (BESA box) and fix the light to the box with M3.5 machine screws.

■ For other spacings, mount the fitting over a narrow *architrave* box and screw to the wall on each side using woodscrews and wall fixings.

The fitting may have a terminal block for connecting the spur cable. If there is just a flex, connect this to the cable via a separate 2 or 5 amp terminal connector block. There may be a separate heat resistant sheathing which must be slipped over the flex.

LSC plug bases slot into a matching wall socket mounted on a special recessed wall box. Connect the spur cable to the socket as shown below.

Earth connections

Some lights are double-insulated and don't have an earth terminal. What you do with the cable earth then depends on the type of enclosure.

■ For a metal architrave box or LSC wall socket, sleeve the spur cable earth core and connect it to the earth terminal on this.

■ For a plastic conduit box, cut back the bare core where it emerges from the sheathing.

Fitting procedure

Wall lights are generally best placed 1.5m (5') above the floor, though bedhead lights should be fitted around 300m (12") lower down.

■ Mark out the cable run down (or up) to each wall light position, and cut out a chase for the cable. You can run the cable in conduit, or simply secure it in the chase with clips and then plaster over it.

■ At the light position, cut a recess to take your conduit box, architrave box or LSC wall box. Set it in position, flush with the plaster, and feed in the cable.

■ Make good around the enclosure.

■ Prepare and connect the cable cores to the light fitting, then mount the fitting on the wall.

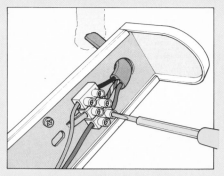

Enclosed base fittings usually have connector blocks inside the base for connecting the spur cable. Don't forget to sleeve the earth core . . .

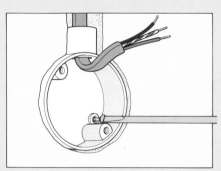

Open base fittings with 51mm (2") hole spacings can go on a conduit box. Mount this in the wall, feed the cable through, and make good around the hole . . .

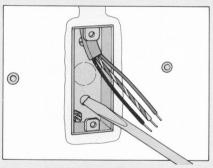

Non-standard open base fittings go on a narrow architrave box. Mount this in the wall, and make separate wall fixings for the fitting itself . . .

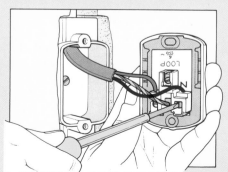

LSC fittings plug into a separate socket mounted in a special wall box. Recess the box in the wall and connect the spur cable to the socket terminals . . .

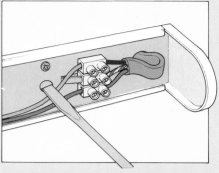

. . . **then simply reassemble** the fitting, mark the wall fixing positions, and screw to the wall using wallplugs or cavity fixings as appropriate.

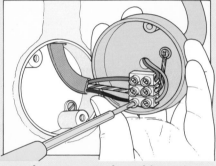

. . . **then connect the cable** to the fitting; the one shown here has a connector block on the base. Afterwards, fit the light to the conduit box using machine screws.

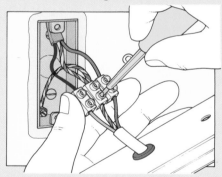

. . . **then connect up;** this fitting needs a separate terminal block to join its flex to the spur cable. Afterwards, screw the fitting to the wall over the box.

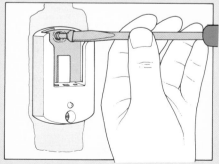

. . . **then fit the socket** in the wall box. The light itself simply plugs into it, allowing easy removal for cleaning or when redecorating the walls.

FITTING SPOTS, TRACK AND DOWNLIGHTERS

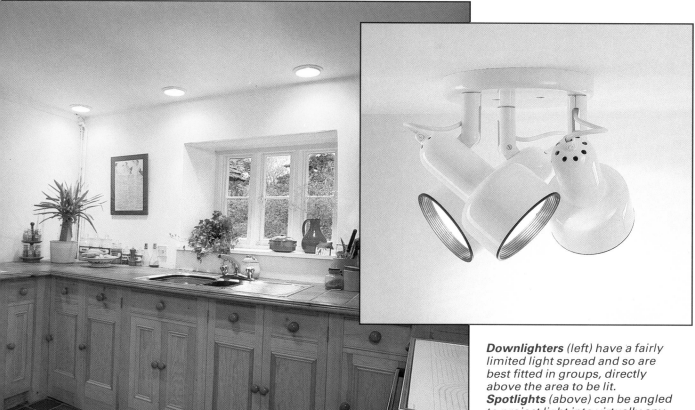

Downlighters (left) have a fairly limited light spread and so are best fitted in groups, directly above the area to be lit.
Spotlights (above) can be angled to project light into virtually any part of the room, so the mounting position isn't critical.

Modern light fittings can transform your decorations – and with careful setting up will give either soft, glare-free background lighting, or punchy highlights as required.

There are three main alternatives to pendants and chandeliers:
Spotlights are inexpensive and come in single, double and triple configurations. If you fit a spotlight in place of the existing ceiling rose, it may be as easy to install as a pendant light. The main factor affecting this is the type of wiring connections.

Wiring Regulations insist that all connections are made inside an enclosed heatproof box. If the spotlight has an enclosed base with the terminals inside, it should meet this requirement so you can screw it directly to the ceiling and lead the cables inside. Lights without an enclosed baseplate must be mounted over a plastic conduit box (BESA box) recessed into the plaster. If a conduit box isn't already fitted it could mean a lot of extra work to install one.

The other possible complication is if you want a spotlight in a different position from the existing rose. This will mean extending the wiring (see Problem Solver).
Track lighting consists of a long conducting track into which lights can be plugged wherever you need them. This is quite expensive but easy to fit and wire.

Wiring is normally straightforward; because the track takes power where you need it you are less likely to have to extend the existing cables. Tracks can be tailored to suit small or large rooms by adding extra lengths with purpose-designed couplers.
Downlighters are often more difficult to fit; because they are intended to be positioned over the area to be lit they usually require extra wiring (see Problem Solver). Recessed types need a large hole cut in the ceiling (making them unsuitable for lath and plaster) and a fair amount of clearance above. Surface-mounted downlighters can be fitted to lath and plaster ceilings.

.... Shopping List ....

The extra fittings you need depend mainly on the system you choose:
Spotlights Buy a conduit box and terminal block if the connections are exposed.
Track lighting Track comes in 1–3m (3–10') lengths:
Track couplers in various shapes are used to join sections.
Lampholders (the individual lights) must be compatible with the track system. Pendant lights and other standard accessories can be connected using a *track adaptor*.
Downlighters normally need no extras except to extend the wiring. Check your ceiling is plasterboard before buying a recessed type.
To extend the wiring buy 1mm^2 two-core and earth cable, green/yellow earth sleeving and a junction box (3-terminal for junction box lighting, 4-terminal for loop-in).
Tools checklist: Drill and bits, screwdrivers (standard and electrical), wire strippers and possibly a padsaw.

FITTING SPOTLIGHTS

In a small room, a spotlight cluster is normally fitted in the centre of the ceiling, in place of an existing rose. This makes the connections simple, but if the light needs to go elsewhere see Problem Solver on extending the wiring. Where the light has an enclosed base, make the connections inside it – otherwise fit a conduit box recessed into the surface of the ceiling.

Lightweight spotlights can be hung from the ceiling plaster using hollow wall fixings, but heavier ones should be screwed to a joist. If the chosen position for the new light doesn't coincide with a joist, the only solution is to fit a mounting board between two joists and screw the light to it.

Fitting a conduit box to take a spotlight also involves fitting a mounting board between the joists. The light then fixes directly to the box – boxes have screw threads to take M4 (4mm) machine screws which may come with the light.

Make certain the power is off before you start. Switch off at the consumer unit and remove the fuse or MCB for the circuit you are working on. Flick the light switch to check.

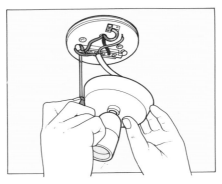
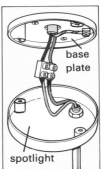

1 With the power switched off, remove the existing rose. If there is more than one cable present, label them to show the connections.

2 Some spotlights have a two-part base which encloses the connections. If the base is open, you must fit a conduit box and wire up inside it.

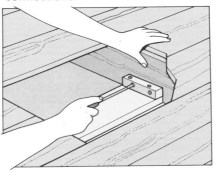
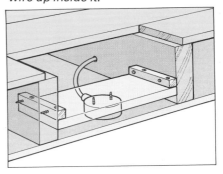

3 To fix a heavy spotlight between joists you need to fit a mounting board to take the screws. Lift floorboards to gain access from above.

4 If you need to fit a conduit box, fix a mounting board so that you can screw it in place flush with the ceiling. Lead the cables into the box.

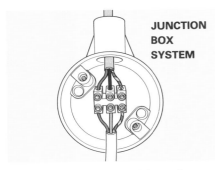

JUNCTION BOX SYSTEM

5 *Wire to a single cable* so that red goes to Live, black goes to Neutral and green/yellow goes to Earth. Use a terminal block connector if none is provided.

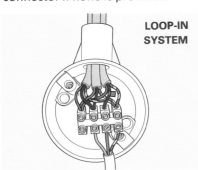

LOOP-IN SYSTEM

6 *For multiple cables* (loop-in wiring) wire up using 15 amp terminal block connectors to restore the connections in the same order as in the rose.

7 If you fitted a conduit box, attach the light to it by using machine screws. If the light has a two-part base, the machine screws are used to attach the light to the base after this has been screwed in place with woodscrews.

INSTALLING A TRACK SYSTEM

Track fixings must be screwed into the joists, so work out the track position before you start.

Flush fitting tracks are screwed straight to the ceiling and normally must run under a joist.

Clip fixing tracks are attached to metal clips screwed to the joists. The clips may be positioned at any point on the track, so the track can run with or across the joists.

Suspended fixings (brackets or wires) allow the track to hang away from the ceiling. Like clip fixings they run with or across the joists.

Switch off at the consumer unit before wiring into the lighting circuit. There are two options:

■ The easy option (the only one for suspended track) is to run a flex from the existing rose to the track terminals. On a loop-in rose this avoids having to fit a junction box to take all the connections.

■ The alternative for a flush fitted track is to remove the rose and wire directly to the lighting cables. On a loop-in system, or if the cable is too short, restore the connections using a junction box and add an extension cable (see Problem Solver). Patch the hole left by the rose or fit a screw-on *rose cover*.

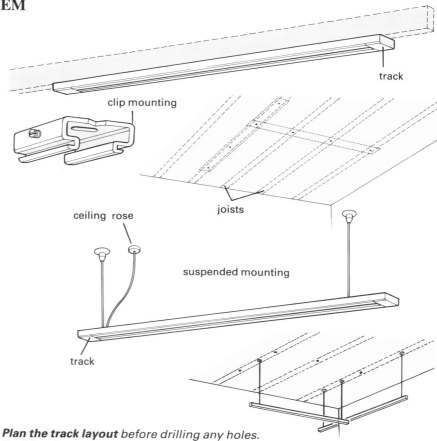

Plan the track layout before drilling any holes. Decide where to connect into the lighting circuit, and align the track so that the mountings are screwed into the joists.

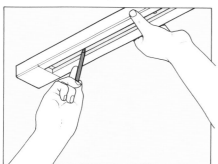

For a direct fixing system position the track with the cable entry hole over the ceiling rose position, mark the screw positions and drill the holes.

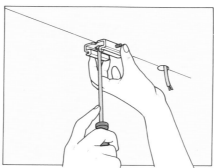

For clip fixing systems mark the track position with a pencil. Position the clips to align with the joists, make the screw holes and screw the clips in place.

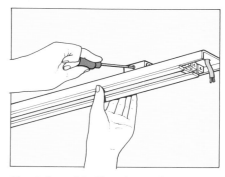

Feed the cable/flex through the entry hole. Screw the track to the ceiling, or slot it over the mounting clips, then tighten the screws to secure the track.

To join tracks slide in the connector and lock it in place with the screws provided. Slide in the next track section, lock it, and tighten all the fixings.

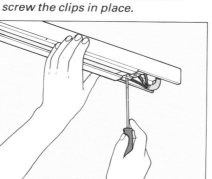

Wiring up depends on whether you wire to the cable or flex. Connect red or brown to live, black or blue to neutral, and green/yellow to earth.

Trade tip

Patching the holes

❝ When making good the hole after removing the ceiling rose, stuff a wad of paper into the hole. This gives the filler something to grip on and stops it pushing through the plaster.

For larger holes, use a scrap of expanded polystyrene.

Apply filler over the top and leave to harden thoroughly before sanding. ❞

FITTING A DOWNLIGHTER

Surface-mounted downlighters are fitted like spots. All other types go into holes cut in the ceiling, and the size of the hole is critical – if it's too loose the fixing clips won't grip. Most lights come with a template for accurate marking.

You are unlikely to want a downlighter in the same place as the old light so the wiring is likely to need extending (see Problem Solver). With most types the cable leads straight into the back and a junction box can be used to take the original rose connections. Before wiring up, make sure the power is off at the consumer unit.

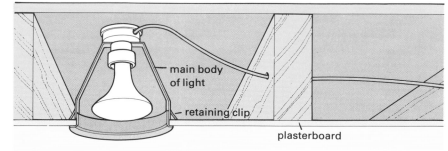

main body of light

retaining clip

plasterboard

outer flange

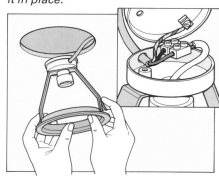

Recessed downlighters fit into a hole in the ceiling. When the light is pushed up, the legs of the clip spring outwards to lock it in place.

1 *Mark the hole using a template or the light itself. Drill a hole inside the line then use a padsaw to cut out the circle. Be careful not to cut it too large.*

2 *The downlighter should be a snug fit in the hole. If it's too tight, ease the sides gently with coarse sandpaper until it fits properly.*

3 *Connect the cable to the light. Adjust the retaining clip to suit the thickness of the ceiling then push the downlighter into place until it locks in.*

PROBLEM SOLVER

Extending the wiring

To extend the lights to a new position use $1mm^2$ cable.

If you can work from above drill a hole 50mm (2″) below the top of each joist, midway between the floorboard nails, to run the cable across them.

If there is no access from above, cut small holes under each joist to pull the wiring through. Probe to find the joist positions, and cut a 100×50mm (4×2″) hole in the plaster below each one. Notch each joist to take the cable using a chisel.

Make a loop in the end of the new cable and feed it towards the first hole starting from the existing wiring. Hook the cable from each hole in turn using a bent coathanger wire. Screw metal plates to each joist to hold the wire in the notches. Finally, patch the ceiling with plasterboard and filler.

Connect up with a junction box taking the place of the existing rose and joining the new cable.

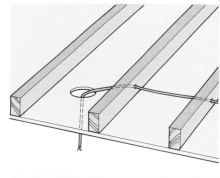

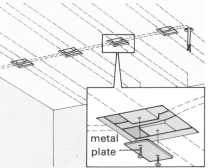

metal plate

Run cable through holes in the joists by lifting floorboards, or run it underneath the joists, using a bent coathanger to pull it through holes cut in the plaster.

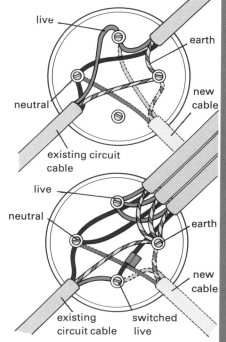

live

neutral

earth

new cable

existing circuit cable

live

neutral

earth

new cable

existing circuit cable

switched live

Wire up to the existing lighting circuit using a junction box to make the same connections as in the rose, with the new cable in place of the original flex.

FITTING DIMMER SWITCHES

Dimmer switches allow you to set the light level anywhere between off and full on, and come in styles to blend with any room. The controls vary (see below), but models with separate on/off switching can be left at a preset level, while others have to be dimmed right down to switch off.

Choosing a dimmer

Apart from looks and convenience, there are several practical points to check when choosing:

■ Dimmers can be used with all types of light fitting, but most only work with tungsten filament bulbs (fluorescent tubes need special dimmers, which are not covered here). If you want the dimmer to operate a large number of powerful bulbs, or one very low power one, check that its power rating is suitable.

If you want to fit dimmers to lights with two-way switching (such as hall lights controlled from upstairs and down), not all types are suitable. Two-way dimmers may be wired in pairs like conventional switches, but you can also get 'master' switches with special exten-

Dimmer switches are perfect for controlling harsh overhead light.

sion controls to allow switching from several points.

■ Dimmers normally fit straight into the old switch mounting box. However, if you are replacing a double (*two-gang*) switch or other multiple switch, your choice is more limited unless you replace the

mounting box. Double switches are usually the same width as single ones, but most double dimmers are designed for wider boxes.

■ If your existing switch is fitted in a shallow mounting box, this may also affect your choice unless you replace the box too.

....Shopping List....

single gang rotary dimmer

'Georgian' brass rotary dimmer

'hi-tech' rotary dimmer

rotary dimmer with rocker switch

single gang rotary dimmer

single gang rotary dimmer

touch dimmers

remote control dimmer

remote control unit

two-gang rotary dimmer (120mm fitting)

two-gang rotary dimmer

two-gang touch dimmer

Dimmers come in styles from traditional to hi-tech, with various types of control:

Rotary dimmers are most common, and lowest in price. They can have the on/off control combined as a rotary switch or there may be a separate rocker switch.

Touch switches turn on and off with a light tap on the faceplate and are controlled by holding your finger down. They often have extra extension controls.

Remote control dimmers come with a hand-held transmitter which beams a signal to the switch. They can also be operated by touch.

Multi-gang dimmers are available with rotary or touch operation and two, three or four controls. Most need wide boxes.

The only tool normally required is a small electrical screwdriver. If you change the mounting box, you may need extra tools as well as the new box itself (see overleaf).

CHANGING A SWITCH TO A DIMMER

Check your existing switch as shown (right) and described on page 73. If it is a multi-gang or two-way switch you have much more limited choice of replacement. Also check the depth of the mounting box. If this is too shallow it must be replaced at the same time.

Surface mounting boxes can be measured without undoing anything. To check a flush mounted box, switch off the power and undo the screws, then ease off the faceplate.

To fit the new dimmer, turn off the power, unscrew the existing switch and remove the wires from their terminals after checking which one goes where and labelling them.

If the mounting box needs replacing, do this next. Then, unless otherwise instructed, connect the wires to the dimmer in the same way as the original switch.

Turn off the power

Before you start, turn off the main switch at the consumer unit (fusebox). If you need power in the meantime, remove the fuse (or trip the MCB) for the circuit of which the light is a part and turn on the main switch again. Test the light switch to double-check the circuit is dead before proceeding.

TYPICAL OLD AND NEW CONNECTIONS

existing switch
(one way, single gang)

dimmer switch

two-way switch   multi-gang switch

multi-gang dimmer

Single-gang switches *normally have two terminals and the wires can go either way round. If there are three terminals to allow two-way switching, one is usually labelled 'Common' and the others something like '1' and '2'. If this type is wired for one-way switching, the extra terminal is left empty, so one wire will go to Common and one to 1.*
Two-way switches *have a third wire to the spare terminal.*
Multi-gang switches *have groups of terminals like single switches.*

Dimmer connections *are often the same as an ordinary switch, but sometimes there are extra terminals for two-way switching and extension controls. There should be an instruction sheet showing which terminals correspond to the terminals in the switch you are replacing.*

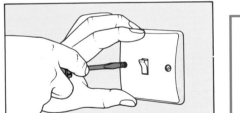

1 *With the power off at the consumer unit, undo the screws holding the old switch plate and ease it away. Unscrew the terminals and remove the wires.*

Trade tip

A switch in time . . .

❛ *If you are replacing a switch which is connected with more than two wires, label them so you can remember which ones went where. Use a piece of masking tape and write the name of the terminal on it before you remove the wire. This makes it easy to find the corresponding connections on the dimmer.* ❜

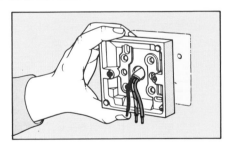

2 *If you need to replace a surface mounting box, undo its fixing screws and lift it away from the wall. Ease it over the projecting cable.*

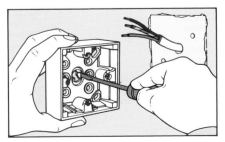

3 *Knock out one of the blank holes in the new box so you can fit it over the cable. Screw it to the wall, drilling and plugging new holes if necessary.*

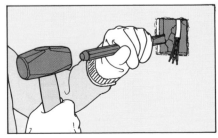

4 *To replace a flush box you need to chop out a deeper hole with a cold chisel. Thread the cable through a knock-out and screw the box to the wall.*

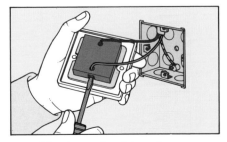

5 *Slip the ends of the wires into the terminals on the new switch and tighten the screws. Then fit the faceplate to the box and secure the screws.*

FITTING SECURITY LIGHTING

Many break-ins happen at night, or when it's obvious that there's no one at home. If you're just going out for an evening, you can leave on a light or two. But for longer periods such as holidays, a more effective deterrent is to fit a device that switches the lights on and off and gives the impression that there's always someone in.

What's available

Providing your home with security lighting need be neither costly nor difficult. There are a number of different devices available (see chart), the majority of which are simple plug-in timers, or light-sensitive controls that fit in place of an existing light switch.

For more comprehensive protection, consider swapping an ordinary outdoor light for a replacement infra-red detector light, or wiring an infra-red control unit into the existing fitting. Installing these is more involved, since both must be wired into the outdoor lighting circuit. But since they react to human presence, rather than preset programmes or light levels, they are a much more convincing deterrent.

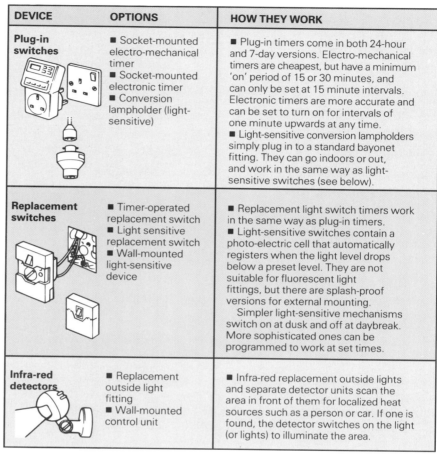

DEVICE	OPTIONS	HOW THEY WORK
Plug-in switches	■ Socket-mounted electro-mechanical timer ■ Socket-mounted electronic timer ■ Conversion lampholder (light-sensitive)	■ Plug-in timers come in both 24-hour and 7-day versions. Electro-mechanical timers are cheapest, but have a minimum 'on' period of 15 or 30 minutes, and can only be set at 15 minute intervals. Electronic timers are more accurate and can be set to turn on for intervals of one minute upwards at any time. ■ Light-sensitive conversion lampholders simply plug in to a standard bayonet fitting. They can go indoors or out, and work in the same way as light-sensitive switches (see below).
Replacement switches	■ Timer-operated replacement switch ■ Light sensitive replacement switch ■ Wall-mounted light-sensitive device	■ Replacement light switch timers work in the same way as plug-in timers. ■ Light-sensitive switches contain a photo-electric cell that automatically registers when the light level drops below a preset level. They are not suitable for fluorescent light fittings, but there are splash-proof versions for external mounting. Simpler light-sensitive mechanisms switch on at dusk and off at daybreak. More sophisticated ones can be programmed to work at set times.
Infra-red detectors	■ Replacement outside light fitting ■ Wall-mounted control unit	■ Infra-red replacement outside lights and separate detector units scan the area in front of them for localized heat sources such as a person or car. If one is found, the detector switches on the light (or lights) to illuminate the area.

A well lit up house is far less likely to attract the attention of burglars than one which looks dark and empty. With the right switching devices you can programme the lights to mimic occupation with startling realism.

PLUG-IN SWITCHES

Timers plug straight into a socket outlet. They are rated at 13 amps, so you can power more than one light – and perhaps a radio too – to make the effect more realistic.

Light-sensitive bayonet-fit switches plug into any standard bayonet lampholder.

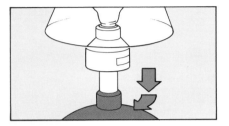

Bayonet-fit switches plug into any bayonet lampholder. Make sure that the sensor 'window' is towards the light and that it isn't masked by a lampshade.

Electro-mechanical timers (above) have levers on the dial for time setting. Once set, simply plug into a socket.
Electronic timers (inset) have push-button controls and digital display.

REPLACEMENT SWITCHES

Replacement switches screw to the wall in place of the existing light switch. All switches, whether timer controlled or light-sensitive, will work with both one- and two-way lighting circuits. If the existing faceplate has two switches controlling two different lighting circuits, make sure the replacement one is a double version too. This won't make both circuits into security lights – only one of the two switches has this option.

■ First switch off the power at the consumer unit.
■ Undo the two screws and ease the switch carefully from the wall.
■ Release the wires from the switch (leave the earth attached to the backing box) and transfer them to the new switch following the wiring below for a one- or two-way circuit.
■ Screw the switch to the box.

On timer-operated switches you may have a choice of setting functions. All designs allow you to set the on/off timings, but on some the switch can be set to programme itself by memorizing and repeating one day's on/off pattern.

ONE-WAY CIRCUIT

circuit cables

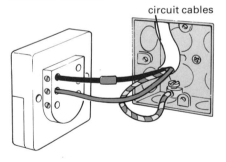

On a one-way circuit, black goes to the terminal marked 'on/off' or L1, red to 'common' or L2. The 'two-way only' or L3 terminal isn't used.

TWO-WAY CIRCUIT

circuit cables

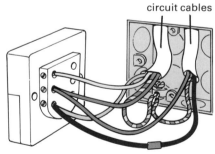

On a two-way circuit, yellow goes to 'on/off' or L1, the two reds go to 'common' or L2, and black and blue go to 'two-way only' or L3.

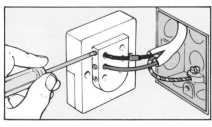

Some wall switches have a built-in timer that turns on the light a set time after it gets dark. The adjuster is neatly concealed under a sliding cover.

To fit a replacement switch, unscrew the old one and remove the wires. Reconnect the wires to the new switch and screw it to the existing backing box.

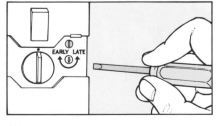

When dusk falls, set a light-sensitive switch to 'automatic' then adjust the light level sensitivity control until the light just comes on.

INFRA-RED REPLACEMENT LIGHTS

Infra-red replacement outdoor lights come in two types: an enclosed wall light with the detector built-in, or a floodlight unit with the detector mounted on the same baseplate. On some floodlight designs the detector can be removed from the baseplate and mounted independently.

All-in-one units are the easier of the two to fit, but separate units are a little more versatile when it comes to setting up the system (see below). Some kits have the facility to operate extra lights up to a maximum of around 500 watts. Use one of these to operate a pair of floodlights for the garden, leaving the light itself to cover the house.

Replacement units simply screw to the wall in place of the existing light and connect to the wiring that's already there. Be sure to turn off the power before you start.

1 *Remove the old light from the wall and disconnect the wiring. Offer up the baseplate of the new light and mark the fixing hole positions on the wall.*

2 *Drill and plug the fixing holes. Screw the light to the wall and connect the new light – red to L, black to N and earth to the earth terminal.*

3 *If there isn't a sealing gasket between the light and the wall, apply mastic sealant to the light baseplate. Finally, secure the cover to the baseplate.*

SETTING UP THE DETECTION AREA

Whether fitting a replacement light, or an add-on detector (see overleaf), the positioning of the unit is crucial. You don't want the light switching on every time someone walks by.

Most security lights come with detailed diagrams showing how the coverage area varies depending on the height of the sensor above the ground. The ideal height is around 2–2.5m (6′6″–8′), with the head tilted down at about 15°.

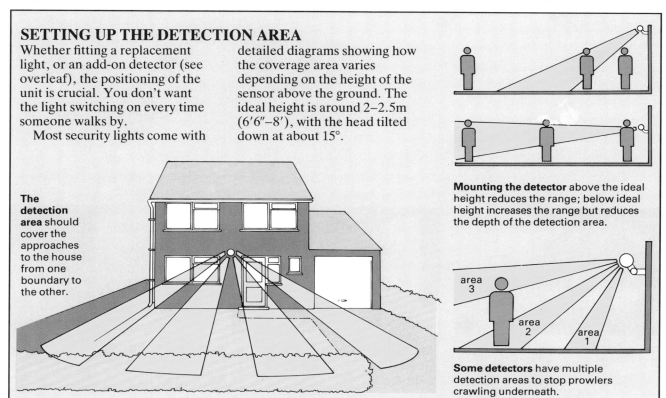

The detection area should cover the approaches to the house from one boundary to the other.

Mounting the detector above the ideal height reduces the range; below ideal height increases the range but reduces the depth of the detection area.

area 3
area 2
area 1

Some detectors have multiple detection areas to stop prowlers crawling underneath.

FITTING AN ADD-ON CONTROL UNIT

An add-on infra-red detector converts any existing outside light to infra-red operation. The detector mounts on an outside wall and connects into the circuit between the light switch and the light. It can operate any number of lights so long as the total load doesn't exceed its design capacity – usually around 1000 watts.

There are two wiring options. **Automatic operation** is the easiest to arrange but stops you leaving the light on permanently. Simply run a new length of 1mm^2 three-core and earth cable from the detector to the existing light and wire into it using a block connector as shown below. All connections must be made inside the light fitting.

Make sure you turn off the power at the consumer unit before you start. After fitting, test following the manufacturer's instructions. **Manual operation** allows you to override the automatic function and use the light as normal, but is much more difficult to set up. The switch indoors must be changed for a double one with extra wiring to the light – a job for an electrician.

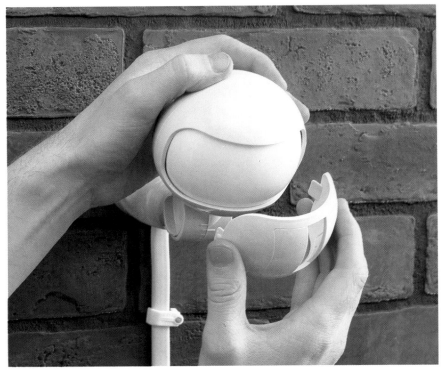

Fit a separate add-on detector so that it has an unobstructed view and adjust it horizontally and vertically to give the best detection area. In a small garden, narrow the area using the masking screens supplied in the kit to avoid false alarms.

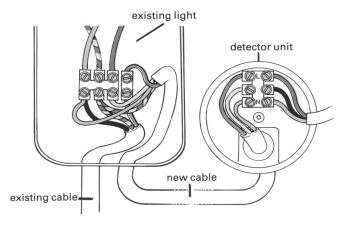

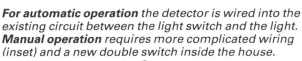

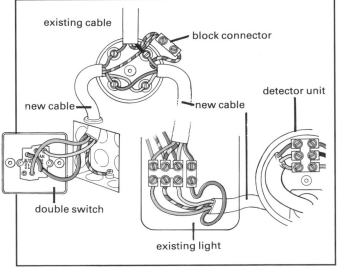

For automatic operation the detector is wired into the existing circuit between the light switch and the light. *Manual operation* requires more complicated wiring (inset) and a new double switch inside the house.

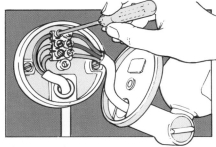

Connect the new cable to the detector. Red goes to the L terminal, blue to N, and yellow to the other terminal – which may be marked 'switched live'.

Having mounted the detector, run the cable to the existing light using cable clips. Unscrew the light from the wall to reveal the existing wiring.

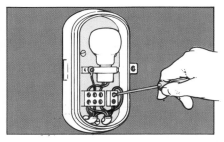

Disconnect the existing red wire and link to the new cable's red via a block connector. Connect the new blue to the N terminal, and the yellow to L.

INDEX

ACKNOWLEDGEMENTS

Photographers

Arcaid (Richard Bryant) 2 (Ken Kirkwood) 1(tl); Barnabys 10; Collins 15-18; Crown 63; Eaglemoss (Jon Bouchier) 11, 49(tl), 49(b), 73(t), (John Suett) 25, 49(tr), 53(b), (Derek St Romaine) 21-22, (Steve Tanner) front cover(tr), 23, 27-30, 35, 37, 70, 73(b), 78; Electricity Council 7; Forbo Nairn 61; Robert Harding Picture Library front cover(tl), 1(br); Smiths 76(b), 77; Stag 53(t); Superswitch 75, 76(t); Elizabeth Whiting Associates 65, 69, (Michael Dunne) front cover(br), (Rodney Hyatt) 4, (Spike Powell) 6.

Illustrators

Peter Bull 74; Neil Bulpitt 23-24, 25-26, 31-34; Kuo Kang Chen 19-22, 65-68; Paul Cooper 11-14; Jeremy Dawkins 8-9, 27-30; Paul Emra 50-52, 63-64; Jeremy Gower 50-52; Andrew Green front cover(bl), 11-14, 19-22, 27-30, 31-34, 35-38, 39-42, 53-56, 57-62; Maltings Partnership 75-78; Stan North 43-46, 53-56; Colin Salmon 11-14; Peter Serjeant 15-18.